Food Microbiology and Food Safety

Food Microbiology and Food Safety publishes valuable, practical, and timely resources for professionals and researchers working on microbiological topics associated with foods as well as food safety issues and problems.

The series is co-published with the IAFP. The International Association for Food Protection is a non-profit association of food safety professionals. Dedicated to the life-long educational needs of its Members, IAFP provides Members with an information network through its two scientific journals (Food Protection Trends and Journal of Food Protection), its educational Annual Meeting, international meetings and symposia along with international interaction between food safety professionals.

More information about this series at http://www.springer.com/series/7131

Tanja Ćirković Veličković
Marija Gavrović-Jankulović

Food Allergens

Biochemistry and Molecular Nutrition

 Springer

Tanja Ćirković Veličković
Department of Biochemistry
Faculty of Chemistry
University of Belgrade
Belgrade
Serbia

Marija Gavrović-Jankulović
Department of Biochemistry
Faculty of Chemistry
University of Belgrade
Belgrade
Serbia

ISBN 978-1-4939-0840-0   ISBN 978-1-4939-0841-7 (eBook)
DOI 10.1007/978-1-4939-0841-7
Springer New York Heidelberg Dordrecht London

Library of Congress Control Number: 2014942308

Printed on acid-free paper

Springer is part of Springer Science+Business Media (www.springer.com)

Dedicated to Sava, Aleksandar, and Nikola

Preface

Food allergy is an abnormal immunological reaction to food proteins, which causes an adverse clinical reaction. It can occur as a consequence of conformational cross-reactivity to respiratory allergens (oral allergy syndrome), or following sensitization via gastrointestinal tract. Almost all known food allergens that sensitize via gastrointestinal system belong to the prolamin and cupin protein superfamilies of allergens. Those are mainly characterized by resistance to heat and digestion. The discrepancy between the vast numbers of proteins we encounter and the limited number of proteins that actually become allergens, have led scientists to investigate what unique features make proteins destined to be allergens. The information gained from these studies has led to an allergy assessment strategy that characterizes the potential allergenicity of biotechnology products prior to their commercialization. Beside allergen structure, new data are emerging on the influence of various other factors on the allergen survival and uptake by the gut immune system and presentation by the antigen-presenting cells, some of them related to the biophysical and immunomodulatory properties of the food matrix, and other to the way we process food. Nevertheless, the built of knowledge on all those various and complex interactions between food components and gut immune system will help us to better understand food allergies and to manufacture safer food.

The most efficient treatment of food allergy is allergen avoidance. Thus, labelling of food for the allergen content is required by European food production regulative. Clear guidelines have been established in the EU regarding the allergenic food labelling according to EC Directive 2000/13/EC and amendments: Directive 2000/1/EC Annex IIIa and the Directive 2007/68/EC. The former included a list of 12 potentially allergenic food (cereals containing gluten, crustaceous, eggs, fish, peanuts, soybeans, milk, nuts, celery, mustard, sesame, and sulphur dioxide) that must always be declared on the label of the foodstuffs present in pre-packaged food traded inside EU. The latter included two more food (lupine and molluscs).

This monograph covers the topics of biochemistry, digestibility, and uptake in the gastrointestinal tract of the most important food allergens causing IgE-mediated food allergy that is believed to be responsible for most immediate-type, food-induced hypersensitive reactions. Currently available tests and strategies for food allergens identification and quantification in food matrices are reviewed in the monograph, as well as links between food processing, food matrix, and immuno-modulatory components of food that can influence food allergy development and onset of allergic reactions.

The authors of the book are professors of Biochemistry at the University of Belgrade, Faculty of Chemistry and also lead researchers of the Center of Excellence for Molecular Food Sciences at the Faculty of Chemistry. The monograph is also intended for students of the courses of Food Biochemistry and Nutrition, as well as Molecular Allergology of the postgraduate studies at the University of Belgrade, Faculty of Chemistry. We also hope that the monograph will be valuable to all those who are involved in elucidation of allergenic structures which contribute to elicitation of food allergy and food scientists involved in design of safer and more functional food products.

Acknowledgment

The authors acknowledge help of the colleagues who read and critically commented on the final version of the manuscript, especially Dr. Dragana Stanić-Vučinić, Dr. Katarina Smiljanić, Dr. Milica Grozdanović, and Dr. Milica Popović.

We would also like to thank our families for their sustained support and love.

In Belgrade, 06.02.2014   Tanja Ćirković Veličković
In Belgrade, 06.02.2014   Marija Gavrović-Jankulović

Contents

Chapter 1
Food Allergy and Gastrointestinal Tract

Contents

Abbreviations

Ag	Antigen
AhR	Aryl hydrocarbon receptor
APC	Antigen-presenting cell
CCD	Cross-reactive carbohydrate determinants
CCR	CC-chemokine receptor
CXCR	CXC-chemokine receptor
CT	Cholera toxin
DBPCFC	Double-blind, placebo controlled, oral food challenge
DC	Dendritic cell
EAST	Enzyme-allergosorbent test
ELISA	Enzyme-linked immunosorbent assay
Foxp3	Forkhead box protein 3
FAO	The Food and Agriculture Organization
GALT	Gut-associated lymphoid tissue
GIT	Gastrointestinal tract
HDM	House dust mite

T. Ćirković Veličković, M. Gavrović-Jankulović, *Food Allergens,*
Food Microbiology and Food Safety, DOI 10.1007/978-1-4939-0841-7_1,
© Springer Science+Business Media New York 2014

IFN-γ	Interferon gamma
IgE	Immunoglobulin E
IEL	Intraepithelial lymphocyte
IPEX	Immunodysregulation polyendocrinopathy enteropathy X-linked syndrome
iTregs	Induced regulatory T cells
JAK	Janus kinase
LAP	Latency-associated peptide
LPDC	CD103(+) dendritic cell of the *lamina propria*
LTP	Lipid transfer proteins
M cells	Microfold cells
MHC	Major histocompatibility complex
MLN	Mesenteric lymph node
nTreg	Natural regulatory T cell
OAS	Oral allergy syndrome
OFC	Oral food challenge
OVA	Ovalbumin;
PAF	Platelet activation factor
PBMC	Peripheral blood mononuclear cell
PDL1	Programmed cell death ligand 1
PP	Peyer's patch
RAST	radio-allergosorbent test
SPT	Skin prick test
SRS-A	Slow reacting substance of anaphylaxis
STAT	Signal transducer and activator of transcription
TCR	T cell receptor
TGF-β	Transforming growth factor beta
Th	T helper lymphocyte
TLR	Toll-like receptor
TNF	Tumour necrosis factor
Treg	T regulatory lymphocyte
TSLP	Thymic stromal lymphopoietin
WHO	World Health Organization

Summary How harmless food protein becomes recognized by the mucosal immune system as an allergen remains an open question. The pathophysiology of food allergy is characterized by a skewed type 2 T helper response to specific food proteins. More data are needed to explain how regulatory mechanisms of the mucosal immune system fail and result in allergic sensitization to dietary antigens. Gut homeostasis and immunity are a complex interplay of innate and adaptive immune responses. The mucosal immune system is the largest reservoir of immune cells in the body and has an extremely difficult task in distinguishing harmless from harmful antigens, former making majority of signals that mucosal immune system encounters. Normal response of the mucosal immune system to a dietary antigen is an oral tolerance, being in a state of anergy, or a regulated suppression of its immune response.

Mesenteric lymph nodes (MLNs) are essential for the induction of oral tolerance, which depends on Foxp3+ T regulatory (Treg) cells. Migration of Foxp3+ Treg cells from the MLNs to the *lamina propria* occurs via gut-homing signals. CD103+ dendritic cells in MLNs drive the differentiation of Treg cells in the presence of transforming growth factor beta and retinoic acid. Major conduits of antigens to intestinal CD103+ dendritic cells are goblet cells. Intestinal antigen-presenting cells occur in a variety of subtypes and may have distinctive functions in mucosal immunity and regulating gut homeostasis. Signals coming from the diet and microbiome can modulate these interactions and influence mucosal immunity.

1.1 Food Allergy

1.1.1 Basic Facts

Food allergy encompasses a range of disorders that result from adverse immune responses to dietary antigens. This group of conditions includes acute, potentially fatal reactions, and a host of chronic diseases that mainly affect the skin and gastrointestinal tract (GIT).

Food allergies can be divided into: (1) immunoglobulin E (IgE)-mediated reactions (or true food allergy), (2) mixed IgE-mediated and non-IgE-mediated disorders, and (3) non-IgE-mediated diseases [1]. IgE-mediated disorders can be classified as either an immediate GI hypersensitivity reaction or an oral allergy syndrome (OAS). The immediate hypersensitivity syndrome is a disorder that typically involves the skin, respiratory tract, GIT, or generalized reactions, i.e. anaphylaxis. In the most of these patients, serum food-specific IgE antibodies can be measured in conjunction with positive skin tests, confirming the IgE-mediated pattern of the reaction.

The OAS is the form of IgE-mediated allergy based on the contact of food antigens with the mucosal surface; it rarely involves other organ systems. Symptoms of these types of food allergy typically occur within minutes to hours of the ingested food.

Non-IgE-mediated GIT diseases are often classified as dietary protein enteropathies. Dietary protein enterocolitis and celiac disease are the most common forms. Celiac disease is characterized by villous atrophy, crypt hyperplasia, increased intraepithelial lymphocytes (IELs), and a mixed inflammatory infiltrate. Dietary protein enterocolitis and enteropathy are typically caused by cow's milk or soy proteins and cause variable small and/or large bowel injury associated with non-specific villous atrophy and inflammation.

The mixed IgE and non-IgE-mediated disorders include the eosinophilic gastroenteropathies: eosinophilic proctocolitis, eosinophilic gastroenteritis, and eosinophilic esophagitis. These diseases are characterized by an infiltration of the GIT with eosinophils with an absence of other inflammatory cells.

1.1.2 IgE-Mediated Food Allergy

Food proteins that are responsible for IgE-mediated adverse reactions to food are known as allergens. Food allergens are proteins, or glycoproteins, polymers of amino acids, usually of moderate molecular weight (5–70 kDa).

In immediate hypersensitivity reactions, symptoms begin to develop within minutes to an hour or so after ingestion of the offending food, involving abnormal responses of the humoral immune system with the formation of allergen-specific IgE antibodies. In delayed hypersensitivity reactions, symptoms do not begin to appear until 24 h or longer after the ingestion of the offending food and involve abnormal responses of the cellular immune system with the development of sensitized T cells [2].

Allergic response, or hypersensitivity, is a two-stage process: the first stage involves sensitization to the offending protein (allergen) and the second stage is the effector phase of food allergy—hypersensitivity reaction mediated by IgE. An outline of all the events encountered during food-induced allergic reactions is given in Fig. 1.1.

In order to initiate an allergic reaction, food allergens have to reach an immune system through the GIT, namely, mucosal surfaces. After primary exposure, food allergens are captured by antigen-presenting cells (APC), especially dendritic cells (DCs) of *lamina propria* in intestine. The allergens are internalized by DCs by receptor-mediated endocytosis process, macropinocytosis, or by incorporation of microvesicles shed from the surface of neighbouring cells, and by their interaction with nanovesicles or exosomes. The allergens are detected by ubiquitin, a 76-residue protein that is highly conserved in all eukaryotes. The selective attachment of ubiquitin to allergens is the initial signal for a targeted protein degradation. These ubiquitinized allergens move to a proteosomal complex and ultimately get degraded to peptide fragments. The degraded peptide fragments are presented by the major histocompatibility complex class-II (MHC-II) and recognized by naïve CD4+ T cells [3].

The T helper cells (Th cells) or CD4+ T cells have been divided into two broad classes Th1 and Th2, based on the type of cytokines they produce. These CD4+ T cells differentiate into Th2 cells in the presence of interleukin 4 (IL-4; Fig. 1.2). The differentiated Th2 cells secrete cytokines IL-4 and IL-13 which induce class switching to IgE.

The critical role of IgE in both the early and the late phases of allergic inflammation is well established. Initiation of IgE synthesis by B cells requires signals from T cells (Fig. 1.3) [4]. Two signals are needed for B cells to make the isotype switch for synthesizing IgE. The B cell activation signal and class switching to IgE are mainly induced by IL-4 and IL-13 (signal 1) and interaction of CD40 on B cells and CD40-ligand (signal 2) on Th2 cells. Both of these signals are transduced via activation of the Janus family tyrosine kinase (JAK)1 and JAK3 which ultimately lead to phosphorylation of the signal transducer and activator of transcription (STAT)6. This interaction activates the deletional switch recombination, and brings into proximity all of the elements of the functional ε-heavy chain [3, 4].

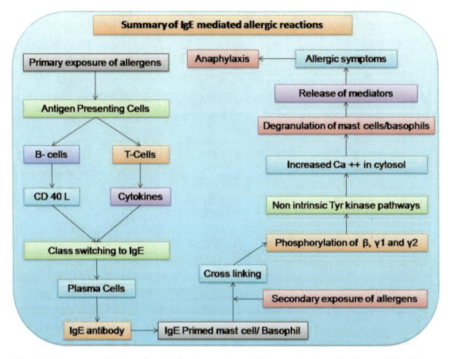

Fig. 1.1 Summary of IgE-mediated allergic reactions. (Figure taken from Reference [3] with permission from Elsevier)

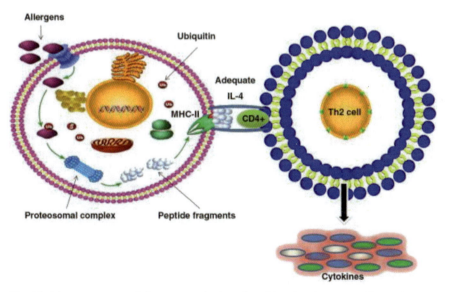

Fig. 1.2 Primary exposure of allergens: production of cytokines by allergen-specific T-cells. (Figures reprinted from Reference [3] with permission from Elsevier)

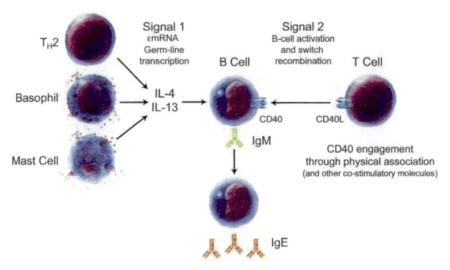

Fig. 1.3 Initiation of IgE synthesis requires signals from cytokines IL-4 and IL-13 and co-stimulation by CD40L. (Figure taken from Reference [4] with permission from Elsevier)

In the first stage of food-allergic reaction development, IgE molecules are formed in predisposed individuals by B cells. IgE molecules through the systemic circulation reach effector cells in allergy: basophils and mast cells (tissue counterparts of basophils). The effector cells in allergy carry high-affinity IgE receptors, which upon cross-linking by an allergen, to which the antibodies were formed (in the sensitization process), release mediators of allergic reaction, such as histamine, prostaglandins, leukotrienes, and other active substances (Fig. 1.4) [5]. Thus, an allergic reaction will activate mast cells and result not only in the release of preformed mast cell inflammatory mediators, such as histamine and tryptase, but also in the synthesis and release of newly generated lipid mediators and cytokines. Hours after the release of allergic reaction mediators, effector cells will start secreting cytokines, IL-4 and IL-13.

Prostaglandins, cytokines, leukotrienes, histamine, slow reacting substance of anaphylaxis (SRS-A), heparin, platelet activation factor (PAF), eosinophil chemotactic factor of anaphylaxis, proteolytic enzymes, and other mediators are secreted by basophils or mast cell's degranulation [3]. These mediators may cause smooth muscle dilation, capillary disruption, local swelling, and other allergic symptoms. In some individuals, these reactions may lead to anaphylaxis or sometimes death.

Histamine acts on histamine 1 (H1) and histamine 2 (H2) receptors to cause contraction of smooth muscles of the airway and GIT; increases vasopermeability and vasodilation; and enhances mucus production, pruritus, cutaneous vasodilation, and gastric acid secretion. Histamine induces IL-16 production by CD8+ cells and airway epithelial cells; IL-16 is an important early chemotactic factor for CD4+ lymphocytes [6]. Tryptase is a major protease released by mast cells and it can cleave complement pathway components C3, C3a, and C5 [7]. Proteoglycans in-

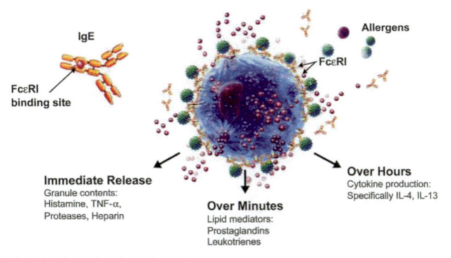

Fig. 1.4 Release of mediators from effector cell after cross-linking of the IgE by an allergen. (Figure taken from Reference [4] with permission from Elsevier)

clude heparin and chondroitin sulphate. Heparin is important for storing the pre-formed proteases and may play a role in the production of α-tryptase. An eosino-philic chemotactic factor of anaphylaxis causes eosinophil chemotaxis, while an inflammatory factor of anaphylaxis results in neutrophil chemotaxis. Eosinophils release eosinophil major basic protein and, together with the activity of neutrophils, can cause significant tissue damage in the later phases of allergic reactions. De-granulation fluids also contain IL-4 and IL-13 that stimulate and maintain Th2 cell proliferation and switch B cells to IgE synthesis. Tumour necrosis factor-α activates neutrophils, increases monocyte chemotaxis, and enhances production of other cy-tokines by T cell [3, 4].

1.1.3   Epidemiology

Adverse immune responses to food affect approximately 5 % of young children and 3–4 % of adults in westernized countries and appear to have increased in preva-lence [8–11]. Food reactions account for one third to one half of anaphylaxis cases in emergency departments in North America, Europe, Asia, and Australia [12]. In addition to gastrointestinal symptoms, food-allergic individuals may experience ur-ticaria, angioedema, atopic dermatitis, oral syndrome, asthma, rhinitis, conjunctivi-tis, hypotension, shock, and cardiac arrhythmias, caused by the massive release of mediators from effector cells, mast cells, and basophiles [13].

Both genetic and environmental factors are important to the development of food allergy. In food-allergic individuals, an IgE is produced against naturally occurring food components, primarily proteins and glycoproteins that usually retain their allergenicity after heating and/or proteolysis. The Food and Agriculture Organization (FAO) and the World Health Organization (WHO) proposed that milk, seashell, egg, fish, peanut, soybean, nut, and wheat are eight major sources of food allergens that cause most of the food allergies. New and emerging food allergens include tropical fruits, sesame seeds, spices, and condiments. These allergies frequently represent a cross-allergy to an allergen derived from another source, e.g. pollens or natural rubber latex, manifested as OAS [14, 15].

1.1.4 Diagnosis and Management of Food Allergy

Tools for diagnosis and management have not changed much in the past two decades, and include the clinical history, physical examination, tests for specific IgE antibody to suspected food, elimination diets, oral food challenges, and provision of medications for emergency treatment [16]. Diagnosis of food allergy is based on history, detailed dietary analysis, skin testing, measuring specific IgE in blood serum, and challenge tests.

For detection of food-specific IgE, in vivo, ex vivo, and in vitro tests are used by allergy specialists. These tests detect sensitisation (the presence of food-specific IgE), but because sensitisation can exist without clinical reactions, the tests generally cannot be used to diagnose food allergy without consideration of the clinical history.

In vivo tests. In vivo detection includes skin prick tests (SPT) and oral food challenges (OFC). For in vivo testing by SPT, the patient should avoid taking medication, i.e. antihistamines for an appropriate length of time prior to testing. With a device such as a lancet, a prick is made through a commercial extract of a food into the epidermis. This allows the test protein to interact with food-specific IgE on the surface of skin mast cells. If the food-specific IgE antibody is present, mast cells degranulate and release mediators that cause localized wheal and flare. Histamine and saline are used as a positive and negative control, respectively, in SPT [16]. SPT is simple and inexpensive, but the wheal sizes can vary according to allergen and subject [10], subjective results can differ between evaluators, and patients with atopic dermatitis may develop false-positive wheals [17]. OFC is considered to be the 'gold standard' in diagnosing food allergy as it can provide more accurate information regarding food allergy. However, these tests are complex, expensive, and time consuming. OFC, particularly for acute IgE-mediated reactions, severe atopic dermatitis, and enterocolitis syndrome, can induce severe reactions. As subject may experience severe adverse reactions, patients who are susceptible to anaphylaxis should not be included in this type of study [17]. Double-blind, placebo-controlled, oral food challenges (DBPCFC) are considered to be the gold standard for diagnosis of food allergy. A recently published position paper includes a summary of methodological issues and gives advice on how to perform DBPCFC [18, 19].

In vitro tests. In vitro studies for determining allergen reactivity include the measurement of serum IgE using radio-allergosorbent tests (RAST), enzyme-allergosorbent tests (EAST), enzyme-linked immunosorbent assay (ELISA), ImmunoCAP assays (Phadia, Uppsala, Sweden), and immunoblotting. By RAST, EAST, and ELISA (competitive inhibition and indirect), multiple samples can be tested at once. However, due to differences with solid phase and sample preparation among analysts, standardization is the main problem. Furthermore, IgG antibodies can compete with IgE antibodies for similar epitopes. Although expensive, Immuno-CAP tests have increased sensitivity compared to RAST, EAST, and ELISA with minimized non-specific binding by non-IgE-binding antibodies. Immunoblotting includes Western blot and dot blot allergen analysis. In Western blotting, proteins are most often tested in their linear conformation so that conformational epitopes may not be represented, and new IgE-binding epitopes, in native conformation hidden within the protein, may be uncovered. Immunoblotting is frequently used since protein bands can be individually analysed to determine the changes in a specific allergen. In dot blot analysis, conformational epitopes may be preserved due to non-denaturing conditions, but in the case of protein mixture, immunogenicity of the entire sample is analysed [11, 16, 20, 21].

Ex vivo tests. By measuring specific IgE, we only measure one interaction, between IgE and an allergen, whilst an allergic response requires two simultaneous interactions of allergen with IgE on the same effector cell. This is simulated in the ex vivo tests based on basophile granulocytes. Ex vivo tests include histamine release or up-regulation of surface molecules CD63 or CD203c on basophile granulocytes, known as the basophile activation test (BAT).

The natural history of food allergy refers to the development of food sensitivities as well as the possible loss of the same food sensitivities over time. Most of the food allergy cases are acquired in the first 1 to 2 years of life, whereas the loss of food allergy is a more variable process, depending on both the individual child and the specific food allergy. For example, whereas most milk allergies are outgrown over time, most allergies to peanuts and tree nuts are never lost. In addition, whereas some children may lose their milk allergy in a matter of months, the process may take as long as 8 or 10 years in other children [22].

Strict avoidance of the offending allergen has been the cornerstone of food allergy management. However, recent studies of egg and milk allergies have presented that some children tolerate egg or milk protein when it is extensively heated [23]. Although the number of studies approaching to treatment of food allergy by oral immunotherapy is increasing, current therapy still relies on allergen avoidance and emergency treatment of severe reactions with antihistamines, corticosteroids, anti-leukotrienes, etc. [24].

Proper labelling of food containing allergens is of crucial importance to help those persons with food adverse reactions. However, food allergens may be hidden and labelling can be non-precise or misleading, thereby hampering prevention. In high-risk infants, there is evidence that exclusive breastfeeding for at least 4 months prevents the development of allergy [25].

From all these reasons, adverse reactions to food, i.e. food allergy and intolerance, have gained considerable attention.

1.1.5 Development of Sensitization to Food Allergens

Antigen degradation in the GIT into tripeptides, dipeptides, and single amino acids allows efficient absorption by enterocytes and results in immune ignorance [26]. Destruction of immunogenic conformational epitopes by the acidic environment of the stomach and by the action of proteases seems to be crucial to promote tolerance via immune ignorance. Fragments of antigens, as small as dipeptides or tripeptides, are ignored by the immune system.

A proportion of partially degraded, or intact antigens, can still reach the surface of epithelium and enter the intestinal mucosa. The ability of molecules to cross the epithelium is dependent on its biochemical properties, such as size, polarity, shape, overall three-dimensional structure, and ability to aggregate into larger complexes [27, 28].

The immune response to dietary antigens in healthy individuals can be tolerant, due to T cell clonal anergy and apoptosis, or active suppression via regulatory T cells with various phenotypes. It is generally accepted that feeding of a high dose of protein antigen leads to tolerance induction, while active suppression takes place after the oral delivery of a low dose of antigen.

This suggests that: (a) the type of the antigen encountered, (b) the proteolytic degradation and intestinal passage, and (c) the dosage of the antigen will influence the type of response subsequently generated in the IgE-mediated food allergy.

The outcome of the immune response will also depend on the genetic background of the individual and can vary from ignorance (no immune response at all) to the induction of antigen-specific antibodies: IgG and secretory IgA.

In genetically susceptible individuals, an inappropriate immune response towards antigen is mounted. Antigen-specific Th2 cells secreting IL-4 and IL-13 favour the production of antigen-specific IgE by B cells. Mast cells (the effector cells of allergic response) are recruited to the *lamina propria* and bind IgE via high-affinity IgE receptor expressed on their surface. That way, the sensitization process to otherwise harmless dietary antigen is completed. In the effector phase of the allergic response, release of mediators from mast cells results in clinical manifestations of food allergy and will further promote permeability of the epithelium to the allergen.

Acute allergic reactions to peanut and tree nuts occur early in life, frequently on the first known exposure to the allergen [29]. Sensitization to peanut proteins may occur in children who have never ingested the peanut through the application of peanut oil to inflamed skin [30]. Allergen-specific levels of IgE and IgA antibodies and their allergen profiles analysed by the allergen chip indicate that IgE antibodies in a cord blood are of foetal origin. Food allergen-specific IgE antibodies were detected more often than inhalant allergen-specific IgE antibodies in cord blood, and the reason for that remains unclarified [31]. Intrauterine sensitization has been suggested to play a role in the development of atopic disease in children, and this has led to current guidelines recommending allergen avoidance during pregnancy.

This chapter focuses on the basics of oral tolerance induction and introduction to the mucosal immunology. Antigen penetration through the epithelium and mechanisms of antigen encountering the immune system are described in Chap. 2. Biochemistry of the most important food allergens causing IgE-mediated food allergy that is believed to be responsible for the majority of immediate-type food-induced hypersensitivity reactions is discussed in Chap. 3.

1.2   Oral Allergy Syndrome

Based on their potential to trigger specific IgE antibody production, food allergens are divided in two classes. The complete or class 1 allergens, besides their ability to cross-link IgE, are also the primary source of sensitization. In general, a majority of class 1 allergens are low molecular weight glycoproteins (70 kDa) with acidic isoelectric points and most of them are highly abundant in food. These proteins are usually resistant to proteases, heat, and denaturants, allowing them to resist degradation during food preparation and digestion [32], thereby enabling direct oral gastrointestinal sensitization [33]. Examples for these class 1 allergens are β-lactoglobulin in cow's milk and stable peanut proteins.

The class 2 or incomplete food allergens are postulated to lack sensitizing capacity. These proteins have the potential to elicit symptoms only after primary sensitization with cross-reactive respiratory allergens and were therefore termed non-sensitizing elicitors. Examples are protein homologues of Bet v 1, the major birch pollen allergen, which are present in fruits and vegetables. Their susceptibility to peptic digestion has been demonstrated and might explain why most often local but not systemic symptoms are triggered on ingestion of Bet v 1 homologues [34, 35]. This phenomenon is known as OAS.

Pollen food syndrome results from cross-reactivity between pollen protein-specific IgE and pollen protein homologous proteins found in fruits and vegetables. These proteins can be grouped into several categories based on their structure and include profilins and pathogenesis-related proteins [36].

Proteins that share common epitopes with Bet v 1, the major birch pollen allergen, occur in pollens of several plant species: apples, stone fruits, celery, carrot, nuts, and soybeans. Two minor allergenic structures—profilins and cross-reactive carbohydrate determinants (CCD) that sensitize approximately 10–20 % of all pollen-allergic patients are also present in a grass pollen and weed pollen. In particular, the clinical relevance of sensitization to CCD is doubtful [37]. Recent reports have demonstrated the biological activity of IgE to cross-reactive carbohydrates in the case of tomato and celery allergies [38, 39] and in the birch–mugwort–celery–spice syndrome [40]. This raised the question of the use of non-glycosylated recombinant proteins in the diagnosis of certain cases of allergic diseases [39].

In patients allergic to fruit and vegetable food, multiple sensitizations to other vegetable products, whether from the same family or taxonomically unrelated, are frequent. The basis of these associations among vegetable food and with pollens lies in the existence of IgE antibodies against panallergens, which determines cross-reactivity [15]. Panallergens are proteins that are spread throughout the vegetable kingdom and are implicated in important biological functions (generally defence), and consequently their sequences and structures are highly conserved. The three best-known groups of panallergens are allergens homologous to Bet v 1, profilins, and lipid transfer proteins (LTP). These proteins are frequently involved in OAS.

Generally, pollen-related allergens tend to be more labile during heating procedures and in the digestive tract compared to allergens from classical allergenic food, such as peanut. Exceptions are three relevant Bet v 1–related food allergens belonging to different botanical families (i.e. Mal d 1 from *Rosacea,* Api g 1, and Dau c 1 from *Apiaceae*). Thermal processing affects the protein structure resulting in abrogated IgE-binding and mediator-releasing capacity, while their potency to induce proliferation in peripheral blood mononuclear cells (PBMCs) or to activate Bet v 1–specific T cells is retained [41].

1.3   Mucosal Immune System

The gut-associated lymphoid tissue (GALT) is the largest reservoir of the immune cells in the body. The physiologic role of the GALT is the ingestion of dietary antigens in a manner that will protect the organism from pathogens, but it will not result in immune reactions to harmless antigens. As such, the GALT is primarily a tolerogenic environment in which a complex interplay of factors creates this property of immunological tolerance [42].

The mucosa of the small intestine alone is estimated to be 300 m^2 in humans, and there are 10^{12} lymphoid cells per metre of human small intestine. Approximately 130–190 g of food proteins are absorbed daily in the gut. The microbiota in the intestine are an additional major source of natural antigenic stimulation and the number of bacteria colonizing the human intestinal mucosa is approximately 10^{12} bacteria per gram of gut content in the colon [43, 44].

There are several distinctive features of the gut immune system that participate in the tolerogenic environment. The inductive sites for immune responses in the gut are Peyer's patches (PPs), isolated lymph follicles located directly within the gut mucosa and mesenteric lymph nodes (MLNs). At these sites, antigen-specific cellular and humoral responses are first generated.

PPs are macroscopic lymphoid aggregates in the submucosa along the length of the small intestine, while MLNs are the largest lymph nodes in the body which serve as a crossroads between the peripheral and mucosal recirculation pathways. M cells overlay PPs and participate in uptake of particulate matter from the gut lumen. MLNs develop distinct from PPs and peripheral lymphoid nodes. Lymphocytes are additionally scattered throughout the epithelium and *lamina propria* of the mucosa [43].

A single layer of columnar epithelial cells separates the gut microflora from the main elements of the gut immune system. Epithelial cells secrete a variety of factors, such as mucin and antimicrobial peptides that contribute to the barrier function. Transport of antibodies, particularly IgA, by epithelial cells also contributes to the barrier function. Below the epithelium lies the mucosa, densely populated with immune cells, including T cells, B cells, eosinophils, and mononuclear phagocytes.

To induce a mucosal immune response, antigen must gain access to APC by penetrating through the mucus layer and then the intestinal epithelial cell barrier [43].

Uptake of antigen occurs through a variety of mechanisms including microfold cells (M cells) associated with PPs and uptake by epithelial cells. In vivo data show that goblet cells (GCs) function as a conduit between lumen and small intestine DCs, allowing passage of antigen [45]. APC and macrophages of the intestinal mucosa are hypo-responsive to many microbial ligands and secrete high levels of IL-10, a cytokine with regulatory function, thus setting up a tolerogenic environment in the mucosa.

IELs, scattered throughout the epithelium and LM, serve to regulate intestinal homeostasis, respond to infection, maintain epithelial barrier function, and regulate adaptive and innate immune responses [42]. The majority of IELs are CD8 + T cells, which express αβ or γδ T cell receptors (TCRs). It has been reported that depletion of γδ T cells impairs induction of oral tolerance.

Thus, the combination of several factors, such as presence of commensals, T cells, and DCs, sets up a tolerogenic environment in the gut. Major factors that condition the gut to be a tolerogenic environment are IL-10, retinoic acid, and transforming growth factor-β (TGF-β) [43, 44, 46, 47].

1.4   Oral Tolerance

Tolerance is the immunologic state defined by a lack of reactivity to an antigen/allergen. In contrast to desensitization, tolerance refers to a permanent immunologic state in which infrequent and repeated antigen exposures do not result in an allergic reaction [46]. Tolerance can be induced, for example, by immunotherapy and is associated with increased regulatory T cell numbers and increased IL-10 production.

Oral tolerance is the state of local and systemic immune unresponsiveness that is induced by oral administration of harmless antigen (i.e. food proteins). Mucosal-induced tolerance appears to be important for the prevention of intestinal disorders such as food allergy, celiac disease, and inflammatory bowel diseases.

A process analogous to oral tolerance also regulates responses to commensal bacteria in the large intestine [43]. There is, however, a difference between tolerance to gut bacteria and tolerance to food proteins: Tolerance to food protein induced via the small intestine affects local and systemic immune responses, while tolerance to gut bacteria in the colon affects local, but not systemic immune responses.

During lymphocyte differentiation, a process of acquiring central tolerance starts with generation of a large repertoire of B and TCRs to all kinds of different antigens including receptors to self-antigens and innocuous foreign antigens, such as food and commensal bacteria. Immune responses against self-antigens are prevented by the process of negative selection of developing T and B cells in the thymus and bone marrow. In addition, tolerance to self-antigens is achieved by natural regulatory T cells (nTregs), defined as T cells recognizing self-antigen with high affinity and expressing the transcription factor forkhead box P3 (Foxp3).

Both processes (central tolerance of T cells and nTreg differentiation) require the interaction of the TCR with its antigen in the thymus. Antigens coming from the intestine are not present or expressed at the sites of lymphocytes differentiation, and for these antigens, additional mechanisms of peripheral tolerance are needed to ensure prevention of deleterious immune responses.

The ability of orally administered antigen to suppress subsequent immune responses, both in the gut and in the systemic immune system, has been referred to as 'oral tolerance'.

The intestine is exposed continuously to huge amounts of foreign antigenic material. As well as ingesting 130–190 g of food proteins per day, the intestine is colonized by commensal microbes (microbiota). The density of these microbes increases along the GIT, reaching up to 10^{12} bacteria per gram of gut content in the colon.

Thus, the intestinal immune system has a complex role in generating protective immunity against harmful antigens and inducing tolerance against harmless materials. A failure in discriminating between harmful and harmless material in the intestine leads to different immunopathologies. Celiac disease and food allergy are the result of intestinal immune system failure to induce tolerance to harmless food proteins [47]. Active immune responses directed against the harmless commensal microbes can result in inflammatory bowel diseases, such as Crohn's disease and ulcerative colitis [48].

Oral tolerance has been demonstrated extensively in rodents using many different model antigens, such as purified proteins, cellular antigens, and small haptens [43]. It has also been shown in humans that oral antigen can effectively modulate subsequently induced systemic antigen-specific immune responses [49]. The effects of oral tolerance are measured typically as reductions in T cell proliferation, systemic delayed-type hypersensitivity, cytokine production, mucosal T cell, and immunoglobulin A responses. Serum antibody responses can also be suppressed, particularly IgE production.

Oral tolerance attenuates a broad range of immune responses and appears to play a central role in immune homeostasis (Fig. 1.5). Its effects on systemic immunity have led to attempts to exploit it therapeutically to prevent or treat autoimmune diseases. Immunomodulation by oral antigen may offer new therapeutic strategies for Th type1-mediated inflammatory diseases and for the development of vaccination strategies.

Studies of oral tolerance to myelin antigens in the experimental autoimmune encephalitis model identified suppressor cells that acted by the secretion of TGF-β. These cells were termed Th3 cells. Since then, the field of active cellular regulation has become a mainstream focus of immunologic investigation. Furthermore, with the identification of Foxp3 as a key transcription factor for thymus-derived nTreg cells, 'suppressor cells' were named 'Tregs', and TGF-β is now recognized as a key cytokine in the induction of Foxp3 + Tregs and other T cell subsets. It has also become clear that GALT is a rich and complex immune network that has evolved to induce immunologic tolerance and Tregs.

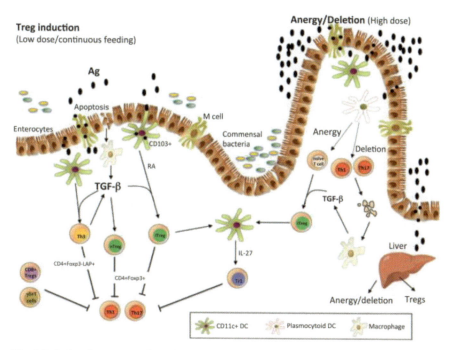

Fig. 1.5 Oral antigen crosses from the intestine into the GALT in a number of ways. It can enter via M cells, be sampled by DC processes that penetrate the lumen, or be taken up by intestinal epithelial cells. DCs in the gut are unique in that they can drive Treg differentiation from Foxp3− cells. These properties of DCs relate to their conditioning by commensal bacteria, TGF-β, and IL-10 from gut epithelial cells, and their expression of retinoic acid, which is provided in the form of vitamin A in the diet and appears to be constitutively expressed by gut DCs. CD11b monocytes may also play a role in the induction of Tregs, and the induction of Tregs occurs in the MLNs and involves both C-C motif receptor 7 (CCR7) and CCR9. Co-stimulation by programmed cell death ligand 1(PDL1) is also important for the induction of Tregs. Macrophages are stimulated to produce TGF-β after uptaking apoptotic epithelial cells or apoptotic T cells following high-dose tolerance. Lower doses of antigen favour the induction of Tregs, whereas higher doses of antigen favour anergy/deletion as a mechanism of tolerance induction. The liver may also play a role in oral tolerance induction and antigen (high dose) may be rapidly taken up by the liver, where it is processed by plasmacytoid DCs that induce anergy/deletion and Tregs. A number of different types of Tregs may be induced or expanded in the gut including CD4+CD25+Foxp3+ iTregs, nTregs, Tr1 cells, LAP + Tregs (Th3 cells), CD8+ Tregs, and γδT cells. TGF, transforming growth factor; RA, retinoic acid; DC, dendritic cells; LAP, latency-associated peptide; Foxp3, forkhead box protein 3; IL, interleukin; MLN, mesenteric lymph nodes. (Figure reprinted from Reference [46] with permission from Wiley)

Oral tolerance can be transferred to a naïve mouse through the transfer of either CD4+ or CD8+ T cells. CD8+ T cells can mediate tolerance, but are not essential to the oral tolerance development. Th3 cells are regulatory CD4+ cells that express latency associate peptide (LAP), but do not express CD25 or Foxp3. Th3 cells are induced by antigen feeding, and their mechanism of suppression is dependent on TGF-β.

Antigen feeding also induces another set of regulatory T cells: CD4+CD25+Foxp3+ cells, termed induced regulatory T cells (iTregs). Foxp3+ regulatory T cells are required for oral tolerance. Treg cells, after their generation in lymph nodes, need to settle in the gut to undergo local expansion and install oral tolerance [50]. Hadis et al. proposed a model of stepwise oral tolerance induction comprising of the Treg cells generation in the gut-draining lymph nodes, followed by migration into the gut and subsequent expansion of Treg cells driven by intestinal macrophages [51].

It has been shown that a lack of Foxp3 leads to the elevated IgE. Immunodysregulation, polyendocrinopathy, enteropathy X-linked syndrome (IPEX) is a rare disease caused by a mutation in the Foxp3 gene leading to an impaired regulatory T cell activity associated with both, skewed Th2 response and auto-reactive phenomena. It has been demonstrated that Foxp3 mutations led to severe food allergy [52, 53]. In children who have outgrown milk allergy, there is a higher frequency of circulating CD4+CD25+ regulatory T cells and decreased in vitro proliferative responses to bovine beta-lactoglobulin in peripheral blood PBMCs compared with children who maintained clinically active allergy, indicating that regulatory T cells may be involved in the development of clinical tolerance to food allergens [54]. In addition, Shreffler et al. demonstrated that a higher frequency of milk allergen-specific Treg cells correlates with a phenotype of mild clinical disease and favourable prognosis [55].

There are multiple mechanisms of oral tolerance, and one of the prime determinants is the dose of the antigen fed. Low doses favour the induction of regulatory T cell, whereas higher doses favour the induction of anergy or deletion. These mechanisms are not exclusive.

The intestinal immune system can be divided into inductive and effector sites. Inductive sites include the GALT, such as PPs, isolated lymphoid follicles, and MLNs; *lamina propria* and epithelium constitute the main effector sites, harbouring large populations of activated T cells and antibody-secreting plasma cells. *Lamina propria* may also contribute to the induction of the tolerance, as a site of antigen uptake and loading of the migratory DCs that encounter naïve T cells in MLNs. Organized structures of the GALT (PPs and isolated lymphoid follicles) seem to be critical for the immune recognition of particulate antigens, such as bacteria and viruses. This reflects the ability of the M cells present in the epithelium of organized structures in the GALT to actively transport material from the gut lumen into the lymphoid areas of the GALT. It has been reported that antigen delivery directly to M cells facilitates induction of tolerance because of a reduction in antigen-specific CD4+ T cells and increased levels of TGF-beta1 and IL-10 producing CD25+CD4+ regulatory T cells in both systemic and mucosal lymphoid tissues [56, 57]. However, other studies reported normal oral tolerance induction in the absence of PPs, demonstrating critical role of MLNs for the induction of high-dose oral tolerance [58, 59]. Thus, it seems that antigen uptake by PPs and isolated lymphoid follicles might play a minor role in oral tolerance to proteins, but may be more important in the process that controls immune responses to commensal bacteria [43].

Antigen uptake by DCs in the *lamina propria* underlying regular villus epithelium seems to be important for tolerance induction to soluble antigens in the small intestine. Despite the presence of proteases and low pH in the upper GIT, some food components are resistant to degradation, and immunogenic material may enter the intestinal lumen. It has been shown that orally administered antigens can be detected in the gut epithelium and *lamina propria* within minutes after feeding the animal. Rapid antigen uptake by DCs in the small intestinal *lamina propria* occurs after feeding mice with ovalbumin (OVA) [28].

The importance of PPs in the induction of mucosal tolerance was investigated in PP-deficient ligated small bowel loops. To explore the requirement for M cells and the PPs in induction of oral tolerance and address the potential in vivo role of intestinal epithelial cells as non-professional APCs, Kraus et al. attempted to induce tolerance in mice with ligated small bowel loops [60]. A small section of vascularized small bowel was spliced away from the gut without disruption of the mesenteric attachments. OVA was then introduced directly into the lumen of the loop prior to footpad immunization. By excising segments of bowel that contain PPs in some mice and segments without patches in others, the authors could study the necessity of the M cell and the underlying patch versus epithelial cells in induction of mucosal tolerance. The results show that OVA-specific T cell proliferation and serum antibody responses are reduced in mice that have previously been given OVA both in loops with and in loops without PPs. Furthermore, both high- and low-dose tolerance could be induced in the absence of PPs. Low-dose tolerance is associated with bystander suppression and requires IL-10, which indicates active suppression and the induction of regulatory cells.

Thus, there is a critical role for components of the mucosal immune system other than PPs in antigen sampling and induction of oral tolerance, and this is consistent with other studies demonstrating that *MLNs are crucial for the induction of oral tolerance* [61]. The transport of antigen from *lamina propria* into the MLNs by CD103+ DCs is the key event for inducing the systemic consequences of oral tolerance. Orally administered antigens are primarily recognized by DCs in the MLN, which require the afferent lymph to process oral antigen, and induction of oral tolerance is impeded by mesenteric lymphadenectomy. The chemokine receptor CCR7, which is necessary for the migration of DCs from the *lamina propria* to MLNs, where CCR7 ligands are expressed, is also important for induction of the oral tolerance. It has been shown that the oral tolerance cannot be induced in CCR7-deficient mice that display impaired migration of DCs from the intestine to the MLNs, suggesting that immunologically relevant antigen is transported in a cell-bound fashion [62]. To investigate the role of DC in regulating the homeostatic balance between mucosal immunity and tolerance, Viney et al. treated mice with fetal liver tyrosine kinase 3 ligand (Flt3L), a growth factor that expands DC in vivo, and assessed subsequent systemic immune responsiveness using mouse models of oral tolerance. Global expansion of DCs exhibited more profound systemic tolerance after the animals were fed with soluble antigen. Most notably, tolerance could be induced in Flt3L-treated mice using very low doses of Ag that were ineffective in control animals. These findings contrast with the generally accepted view of DC as immunostimulatory APC and furthermore suggest a pivotal role for DC during the induction of tolerance following mucosal administration of Ag [63]. Importantly, tolerance can be transferred to naïve animals by transfer of DCs derived from *lamina propria* [64].

DCs can be found in PPs, MLNs, or *lamina propria* of the villus mucosa. All these tissues contain a number of distinctive DC subsets, including some that can preferentially induce the differentiation of regulatory T cells. However, the largest proportion of orally administered proteins is taken up by DC in the *lamina propria* [64]. Intestinal DCs are not inherently tolerogenic. Under physiological conditions, they are capable of presenting antigen and inducing tolerance, but being sufficiently responsive to inflammatory stimuli to allow T cell priming and protective immunity when necessary. Targeting local DCs, either tolerance or active immunity can be achieved [65].

One important feature of oral tolerance to soluble antigens is that it involves the entire organism [43]. One explanation for this could be that orally administered antigens can disseminate systemically via blood and lymph. Liver is a site where this absorbed antigen can contribute to oral tolerance induction. The portal vein drains blood from the intestine to the liver, and injection of the antigen directly into the portal vein is well known to induce antigen-specific tolerance. Food proteins can be detected in the blood of mice and humans soon after eating. Furthermore, serum, as well as exosomes isolated from serum of protein-fed mice, can induce antigen-specific tolerance in naïve recipients [66, 67]. The liver contains several subsets of specialized APC that may contribute to the tolerance induction [68]. Liver sinusoidal endothelial cells efficiently sample circulating antigen, can act as APC, and have been shown to induce tolerance rather than active immunity. Antigen presentation by Kupffer cells and conventional liver DCs also favours tolerance, whereas plasmacytoid DCs in the liver have been implicated in the induction of systemic tolerance to orally administered proteins and haptens. Antigen reaching beyond the liver into peripheral lymph nodes and spleen might be expected to induce tolerance in these sites. Their presentation by resident DCs, in the absence of co-stimulation, will lead to the induction of anergy or Tregs [68].

How luminal antigen gains access to the DCs through the epithelial barrier is still an open question [69]. Materials of low molecular weight, such as haptens and polypeptides, may pass directly across the epithelium by paracellular diffusion through pores in the tight junctions connecting epithelial cells. Conversely, larger molecular complexes can be taken across enterocytes by transcytosis after fluid-phase uptake at the apical membrane. Furthermore, intestinal permeability in food allergic individuals has been shown to be compromised and occurs as a consequence of food allergen exposure, via a mechanism involving active role of mast cells and mediators released upon their activation [47].

1.5   Antigen-Presenting Cells of the Mucosal Immune System

Mononuclear phagocytes generate a lot of research interest. The *lamina propria* underlies the expansive single-cell absorptive villous epithelium and contains a large population of DCs (CD11c + CD11b + MHCII + cells) comprised of two predominant subsets: CD103+CX3+CR1− DCs, which promote IgA production, imprint gut homing on lymphocytes, and induce the development of regulatory T cells, and CD103- CX3+CR1+ DCs, which promote tumour necrosis factor-α (TNF-α) production, colitis, and the development of Th17 T cells. CD11c + cells that express chemokine receptor CX3 + CR1hi could extend dendrites across the epithelial barrier and capture bacteria from the lumen. Recent findings showed that although *lamina propria*-DCs probed the epithelium actively with their dendrites, they did not extend transepithelial dendrites into the intestinal lumen to capture fluorescent antigen in healthy mice [45]. These monocyte-derived cells are more related to tissue macrophages than to tissue DCs; they express surface marker F4/80 and do not constitutively express CCR7, a receptor necessary for DCs migration to the lymph nodes.

CX3+ CR1− CD103+ cells constitutively express CCR7 and migrate to the MLNs, but do not express CX3+ CR1hi and do not extend dendrites across the intestinal epithelium. These cells are derived from common DC precursor in a GM-CSF receptor-dependent manner. CD103 + cells sample antigens from small intestine GCs. GCs function as passages delivering low-molecular weight soluble antigens from the intestinal lumen to the underlying CD103 + *lamina propria*-DCs. The preferential delivery of antigens to DCs with tolerogenic properties implies a key role for this GC function in intestinal immune homeostasis [70]. Antigen passing epithelial barrier by transcytosis, paracellularly or via M cells can also be sampled by these DCs. A special subset of CD103− DCs that express CX3+ CR1int, but do not express surface marker F4/80, was found in the intestine draining lymph, and it is believed that it can be migratory and important for regulating homeostasis in the gut. Recent findings defined a central role for commensals in regulating the migration to the MLNs of CX3CR1(int) mononuclear phagocytes endowed with the ability to capture luminal bacteria, thereby compartmentalizing the intestinal immune response to avoid inflammation [71]. In vivo data revealed that luminal antigen was captured preferentially by CD103+ DCs at a proportion of roughly 10:1 over CD103− DCs and rarely co-localized with plasmacytoid DCs [45].

CD103 + DCs isolated from MLNs preferentially induce generation of gut-homing CD4 + Foxp3 + regulatory T cells from naïve T cells. These cells express Raldh2 which converts retinal to retinoic acid. Both regulatory and gut-homing activities of responder T cells depend on retinoic acid. An important source of the retinal precursor (vitamin A) is diet. CD103 + cells employ various other mechanisms to generate regulatory cells, such as enzyme indoleamine 2,3, deoxygenase, a potent immune-suppressive enzyme, [72] and secretion of regulatory cytokines, such as TGF-βCD103−DCs that express CX3+ CR1int, but do not express surface marker F4/80 are able to induce secretion of interferon gamma (INF-γ and IL-17 from naïve T cells [45, 73], cytokines important in protection against pathogens.

Different subsets of DCs promote development of IgA-secreting, gut-homing plasma cells. DC-derived retinoic acid alone confers gut tropism, but does not promote IgA secretion. However, retinoic acid potently synergizes with IL-6 or IL-5 to induce IgA secretion [74].

Different subsets of intestinal DCs might have distinct roles in regulating mucosal immune system homeostasis and active immunity against pathogens. T cells primed in the MLNs are imprinted to express alpha4beta7-integrin and chemokine receptor CCR9, thereby enabling lymphocytes to migrate to the small intestine. In vitro activation by intestinal DCs instructs expression of these gut-homing molecules. CD103 + DCs have a regulatory function under steady-state conditions, while CD103− DCs may prime the immune response to pathogens. Non-migratory CX-3CR1hi cells express high levels of IL-10 and could expand gut-homing regulatory T cells that were primed in the MLNs. In addition to DCs, stromal MLN cells contribute to the immunoregulatory function by producing retinoic acid and generating gut-homing phenotype in lymphocytes. Mesenteric, but not peripheral, lymph node stroma cells express high levels of retinoic acid-producing enzymes and support the induction of CCR9 on activated T cells in vitro [75].

1.6   Microbiota and Food Allergy

Microbiota are likely to influence development of food allergy, in particular through the modulation of the mucosal immune system [76]. Recent studies showed that signals coming from microbiota can suppress allergic sensitization by influencing IgE production and basophil development [77, 78]. Toll-like receptor 4 (TLR4)-dependent signals provided by the intestinal commensal flora inhibit the development of allergic responses to food antigens [77]. Commensal-derived signals were found to influence basophil development by limiting proliferation of bone marrow resident precursor populations, thus identifying a previously unrecognized pathway through which commensal-derived signals influence basophil haematopoiesis and susceptibility to Th2 cytokine-dependent inflammation and allergic disease [78]. Mice with food allergy exhibit a specific gut microbiota signature capable of transmitting disease susceptibility to germ-free wild-type mice. Disease-associated microbiota may thus play a pathogenic role in food allergy [79]. These animal studies strongly suggest that microbiota-instructed mucosal immunity can influence the development of food allergy. Several human studies reported a possible link between microbiota and food allergy [80, 81]. Furthermore, transplanted healthy infant microbiota mainly composed of *Bifidobacterium* and *Bacteroides* to germ-free mice had a protective effect on sensitization and food allergy development in mice [82].

1.7   Diet-Related Factors that Modulate Immune Response to Dietary Antigens

Obesity is an inflammatory state characterized by the presence of innate Th2 cytokine-producing cells within the adipose tissue and an elevated level of IgE in circulation. Intestinal immune system in obesity is changed, in particular regarding lymphotoxin-induced secretion of IL-23 and IL-22 [83]. Obesity might be a contributor to the increased prevalence of allergic disease in children, particularly food allergy. Systemic inflammation might play a role in the development of allergic disease in obese children [84].

Vitamin A is (retinoic acid), is an essential signal for the development of oral tolerance in MLNs. Although essentially a strong tolerogenic factor, retinoic acid can provide signal to the development of effector Th17 cells and Th1 rather than the regulatory T cells [85, 86]. Data show that in mice, in conjunction with IL-15, a cytokine upregulated in the gut of coeliac disease patients, retinoic acid rapidly activates DCs to induce JNK phosphorylation and release the proinflammatory cytokines IL-12p70 and IL-23. As a result, in a stressed intestinal environment, retinoic acid acted as an adjuvant that promoted rather than prevented inflammatory cellular and humoral responses to dietary antigen [86]. These findings revealed an unexpected role of retinoic acid and IL-15 in the abrogation of tolerance to dietary antigens.

Vitamin D has strong immunomodulatory properties, and it can influence T and B cell homing and suppress development of Th17 cells [87]. It comes from the diet, but it is also synthesized in the skin upon exposure to sunlight. Human studies indicate that vitamin D can have a protective role in food allergy. Interestingly, a protective effect was observed for peanut and shrimp allergy, but not for milk and egg [88]. It is difficult to explain this antigen-specific protective effect of vitamin D in food allergy. Similar to vitamin A, no mechanistic studies in animal models of food allergy have been performed to provide explanation of the mechanism of action and confirm modulatory effect in food allergy.

Aryl hydrocarbon receptor (AhR) is a highly conserved, ligand-inducible transcription factor believed to control the adaptation of multicellular organisms to environmental challenges [89]. AhR ligands come from the diet. Rich sources of AhR ligands are vegetables of the family Brassicaceae, such as broccoli and cabbage. AhR is a crucial regulator in maintaining IEL numbers in both the skin and the intestine [90]. In the intestine, AhR deficiency compromises the maintenance of IELs resulting in heightened immune activation and increased vulnerability to epithelial damage. AhR activity can be regulated by dietary components, such as those present in cruciferous vegetables, providing a mechanistic link between dietary compounds, the intestinal immune system, and the microbiota. Knock-out mice for this receptor have a loss of intestinal $\gamma\delta$ cells and CD8$\alpha\alpha$ T cells. It has been shown in peanut food allergy model that AhR ligands have an immunomodulatory effect [91, 92], but it is not clear whether these ligands can provide protection from food allergy development. Depending on the ligand used, AhR could either promote effector Th17 cells or induce suppression by enhancing regulatory T cells [93].

Polyphenols are abundant in diet and show numerous immunomodulatory effects in both in vitro and in vivo studies. Resveratrol is found in red wine and extensively studied for its effect on the immune system, also in relation to food allergy. Ingestion of resveratrol prevented the development of a food allergy model in mice [94]. Mechanism of action could be inhibition of DC maturation and subsequent early T cell activation and differentiation. Mice fed resveratrol showed reduced allergen-specific serum IgE production and allergen-induced IL-13 and IFN-gamma production from the MLNs and spleens in comparison to the control mice, following oral sensitization with allergen OVA plus adjuvant from cholera toxin (CT). Resveratrol inhibited IL-4, IL-13, and IFN-γ production in splenocytes associated with inhibition of GATA-3 and T-bet expression. Furthermore, resveratrol suppressed CD25 expression and IL-2 production in splenocytes in association with decreased CD80 and CD86 expression levels. Finally, resveratrol suppressed CT-induced cAMP elevation in association with decreases in CD80 and CD86 expression levels in BM-DCs.

1.8   Adjuvants Used in Animal Models of Food Allergy

Important mechanistic data on the development of allergic sensitization to food come from animal studies in mice. Mice do not naturally develop food allergies, but in the presence of orally delivered adjuvants derived from bacteria (CT), Staphyloccocal enterotoxin B, mice develop food allergy to dietary antigen and show increase in antigen-specific IgE and a bias towards Th2 type of immune response. A characteristic feature of the CT adjuvant action is its effect on intestinal DCs. CD103 + cells upregulate OX40L, Jagged 2, and increase secretion of IL-17 and INF-γ [95]. There are no data showing relation of bacterial adjuvants used in development of allergic sensitization in mice to human food allergy.

The epithelium-associated cytokines thymic stromal lymphopoietin (TSLP), IL-25, and IL-33 are suggested to be important for the initiation of allergic responses and Th2-skewing in both respiratory and food allergy. Coupled with the induction of OX40L expression on DCs, the ability of TSLP to limit expression of the p40 subunit suggests that TSLP may indirectly promote a microenvironment permissive for Th2 cell differentiation by limiting the proinflammatory functions of DCs [96].

It has recently been shown that house dust mite (HDM)-induced allergic asthma and food allergy and anaphylaxis to peanut were associated with TSLP production, but developed independently of TSLP, likely because these allergens functionally mimicked TSLP inhibition of IL-12 production and induction of OX40 ligand (OX40L) on DCs. OX40–OX40L interactions are critical for the ability of the DCs to drive Th2 cell differentiation [96]. Blockade of OX40L significantly reduced allergic responses to HDM or peanut [97]. Although IL-25 and IL-33 induced OX40L on DCs in vitro, only IL-33 signalling was necessary for allergic response, probably because of its strong ability to induce OX40L expression on DCs and expend innate lymphoid cells in vivo. IL-33 is a recently identified cytokine member of the IL-1 family. The biological activities of IL-33 are associated with promotion of Th2 and inhibition of Th1/Th17 immune responses. Exogenous IL-33 induces a typical Th2 type immune response in the GIT [98]. Dietary factors that promote IL-33 signalling may likely contribute to the development of allergic sensitization and prompt further research.

In conclusion, it has increasingly being recognized that several factors coming from diet can modulate immune response to dietary antigens. More data are needed to clarify whether their effects are protective or food allergy promoting.

References

1. Eigenmann PA, Beyer K, Wesley Burks A, Lack G, Liacouras CA, Hourihane JO, Sampson HA, Sodergren E: **New visions for food allergy: an iPAC summary and future trends**. *Pediatr Allergy Immunol* 2008, **19 Suppl 19**:26–39.
2. Lemke PJaT, S.L.: **Allergic reactions and food intolerances**. In *"Nutritional Toxicology,"* ed FN Kotsonis, M Mackey, and J Hjelle, Raven Press, NY 1994: pp. 117–137.
3. Kumar S, Verma AK, Das M, Dwivedi PD: **Molecular mechanisms of IgE mediated food allergy**. *Int Immunopharmacol* 2012, **13**(4):432–439.
4. Broide DH: **Molecular and cellular mechanisms of allergic disease**. *J Allergy Clin Immunol* 2001, **108**(2 Suppl):S65–71.
5. Gilfillan AM, Rivera J: **The tyrosine kinase network regulating mast cell activation**. *Immunol Rev* 2009, **228**(1):149–169.
6. Hart PH: **Regulation of the inflammatory response in asthma by mast cell products**. *Immunol Cell Biol* 2001, **79**(2):149–153.
7. Fukuoka Y, Xia HZ, Sanchez-Munoz LB, Dellinger AL, Escribano L, Schwartz LB: **Generation of anaphylatoxins by human beta-tryptase from C3, C4, and C5**. *J Immunol* 2008, **180**(9):6307–6316.
8. Mills EN, Mackie AR, Burney P, Beyer K, Frewer L, Madsen C, Botjes E, Crevel RW, van Ree R: **The prevalence, cost and basis of food allergy across Europe**. *Allergy* 2007, **62**(7):717–722.
9. Gupta RS, Kim JS, Barnathan JA, Amsden LB, Tummala LS, Holl JL: **Food allergy knowledge, attitudes and beliefs: focus groups of parents, physicians and the general public**. *BMC Pediatrics* 2008, **8**:36.
10. Sampson HA: **Update on food allergy**. *J Allergy Clin Immunol* 2004, **113**(5):805-819; quiz 820.
11. Sicherer SH, Sampson HA: **Food allergy**. *J Allergy Clin Immunol* 2010, **125**(2 Suppl 2):S116–125.
12. Sampson HA: **Anaphylaxis and emergency treatment**. *Pediatrics* 2003, **111**(6 Pt 3):1601–1608.
13. Story RE: **Manifestations of food allergy in infants and children**. *Pediatric Annals* 2008, **37**(8):530–535.
14. Hofmann A, Burks AW: **Pollen food syndrome: update on the allergens**. *Curr Allergy Asthma Rep* 2008, **8**(5):413–417.
15. Egger M, Mutschlechner S, Wopfner N, Gadermaier G, Briza P, Ferreira F: **Pollen-food syndromes associated with weed pollinosis: an update from the molecular point of view**. *Allergy* 2006, **61**(4):461–476.
16. Sicherer SH: **Food allergy**. *Lancet* 2002, **360**(9334):701–710.
17. Byrne AM, Malka-Rais J, Burks AW, Fleischer DM: **How do we know when peanut and tree nut allergy have resolved, and how do we keep it resolved?** *Clin Exp Allergy* 2010, **40**(9):1303–1311.

18. Niggemann B, Beyer K: **Pitfalls in double-blind, placebo-controlled oral food challenges.** *Allergy* 2007, **62**(7):729–732.

19. Bindslev-Jensen C, Ballmer-Weber BK, Bengtsson U, Blanco C, Ebner C, Hourihane J, Knulst AC, Moneret-Vautrin DA, Nekam K, Niggemann B *et al*: **Standardization of food challenges in patients with immediate reactions to foods-position paper from the European Academy of Allergology and Clinical Immunology.** *Allergy* 2004, **59**(7):690–697.

20. Burks AW: **Peanut allergy.** *Lancet* 2008, **371**(9623):1538–1546.

21. Stanic-Vucinic D, Stojadinovic M, Atanaskovic-Markovic M, Ognjenovic J, Gronlund H, van Hage M, Lantto R, Sancho AI, Velickovic TC: **Structural changes and allergenic properties of beta-lactoglobulin upon exposure to high-intensity ultrasound.** *Mol Nutr Food Res* 2012, **56**(12):1894–1905.

22. Wood RA: **The natural history of food allergy.** *Pediatrics* 2003, **111**(6 Pt 3):1631–1637.

23. Nowak-Wegrzyn A, Bloom KA, Sicherer SH, Shreffler WG, Noone S, Wanich N, Sampson HA: **Tolerance to extensively heated milk in children with cow's milk allergy.** *J Allergy Clin Immunol* 2008, **122**(2):342–347, 347 e341–342.

24. Kim JS: **Food allergy: diagnosis, treatment, prognosis, and prevention.** *Pediatr Ann* 2008, **37**(8):546–551.

25. Thygarajan A, Burks AW: **American Academy of Pediatrics recommendations on the effects of early nutritional interventions on the development of atopic disease.** *Curr Opin Pediatr* 2008, **20**(6):698–702.

26. Erickson RH, Kim YS: **Digestion and absorption of dietary protein.** *Annu Rev Med* 1990, **41**:133–139.

27. Roth-Walter F, Berin MC, Arnaboldi P, Escalante CR, Dahan S, Rauch J, Jensen-Jarolim E, Mayer L: **Pasteurization of milk proteins promotes allergic sensitization by enhancing uptake through Peyer's patches.** *Allergy* 2008, **63**(7):882–890.

28. Perrier C, Corthesy B: **Gut permeability and food allergies.** *Clin Exp Allergy* 2011, **41**(1):20–28.

29. Sicherer SH, Burks AW, Sampson HA: **Clinical features of acute allergic reactions to peanut and tree nuts in children.** *Pediatrics* 1998, **102**(1):e6.

30. Lack G, Fox D, Northstone K, Golding J: **Factors associated with the development of peanut allergy in childhood.** *N Engl J Med* 2003, **348**(11):977–985.

31. Kamemura N, Tada H, Shimojo N, Morita Y, Kohno Y, Ichioka T, Suzuki K, Kubota K, Hiyoshi M, Kido H: **Intrauterine sensitization of allergen-specific IgE analyzed by a highly sensitive new allergen microarray.** *J Allergy Clin Immunol* 2012, **130**(1):113-121 e112.

32. Astwood JD, Leach JN, Fuchs RL: **Stability of food allergens to digestion in vitro.** *Nat Biotechnol* 1996, **14**(10):1269–1273.

33. Bannon GA: **What makes a food protein an allergen?** *Curr Allergy Asthma Rep* 2004, **4**(1):43–46.

34. Vieths S, Scheurer S, Ballmer-Weber B: **Current understanding of cross-reactivity of food allergens and pollen.** *Ann N Y Acad Sci* 2002, **964**:47–68.

35. Sancho AI, Wangorsch A, Jensen BM, Watson A, Alexeev Y, Johnson PE, Mackie AR, Neubauer A, Reese G, Ballmer-Weber B *et al*: **Responsiveness of the major birch allergen Bet v 1 scaffold to the gastric environment: impact on structure and allergenic activity.** *Mol Nutr Food Res* 2011, **55**(11):1690–1699.

36. Burastero SE: **Pollen-cross allergenicity mediated by panallergens: a clue to the pathogenesis of multiple sensitizations.** *Inflamm Allergy Drug Targets* 2006, **5**(4):203–209.

37. Chardin H, Senechal H, Wal JM, Desvaux FX, Godfrin D, Peltre G: **Characterization of peptidic and carbohydrate cross-reactive determinants in pollen polysensitization.** *Clin Exp Allergy* 2008, **38**(4):680–685.

38. Foetisch K, Westphal S, Lauer I, Retzek M, Altmann F, Kolarich D, Scheurer S, Vieths S: **Biological activity of IgE specific for cross-reactive carbohydrate determinants.** *J Allergy Clin Immunol* 2003, **111**(4):889–896.

39. Westphal S, Kolarich D, Foetisch K, Lauer I, Altmann F, Conti A, Crespo JF, Rodriguez J, Enrique E, Vieths S *et al*: **Molecular characterization and allergenic activity of Lyc**

e 2 (beta-fructofuranosidase), a glycosylated allergen of tomato. *Eur J Biochem* 2003, **270**(6):1327–1337.

40. Bublin M, Radauer C, Wilson IB, Kraft D, Scheiner O, Breiteneder H, Hoffmann-Sommergr-uber K: **Cross-reactive N-glycans of Api g 5, a high molecular weight glycoprotein allergen from celery, are required for immunoglobulin E binding and activation of effector cells from allergic patients.** *Faseb J* 2003, **17**(12):1697–1699.

41. Bohle B, Zwolfer B, Heratizadeh A, Jahn-Schmid B, Antonia YD, Alter M, Keller W, Zuidmeer L, van Ree R, Werfel T *et al*: **Cooking birch pollen-related food: divergent consequences for IgE- and T cell-mediated reactivity in vitro and in vivo.** *J Allergy Clin Immunol* 2006, **118**(1):242–249.

42. Vickery BP, Scurlock AM, Jones SM, Burks AW: **Mechanisms of immune tolerance relevant to food allergy.** *J Allergy Clin Immunol* 2011, **127**(3):576–584; quiz 585–576.

43. Pabst O, Mowat AM: **Oral tolerance to food protein.** *Mucosal Immunol* 2012, **5**(3):232–239.

44. Gourbeyre P, Denery S, Bodinier M: **Probiotics, prebiotics, and synbiotics: impact on the gut immune system and allergic reactions.** *J Leukoc Biol* 2011, **89**(5):685–695.

45. Denning TL, Norris BA, Medina-Contreras O, Manicassamy S, Geem D, Madan R, Karp CL, Pulendran B: **Functional specializations of intestinal dendritic cell and macrophage subsets that control Th17 and regulatory T cell responses are dependent on the T cell/APC ratio, source of mouse strain, and regional localization.** *J Immunol* 2011, **187**(2):733–747.

46. Weiner HL, da Cunha AP, Quintana F, Wu H: **Oral tolerance.** *Immunol Rev* 2011, **241**(1):241–259.

47. Berin MC, Sampson HA: **Mucosal immunology of food allergy.** *Curr Biol* 2013, **23**(9):R389–400.

48. Dupaul-Chicoine J, Dagenais M, Saleh M: **Crosstalk Between the Intestinal Microbiota and the Innate Immune System in Intestinal Homeostasis and Inflammatory Bowel Disease.** *Inflamm Bowel Dis* 2013.

49. Kapp K, Maul J, Hostmann A, Mundt P, Preiss JC, Wenzel A, Thiel A, Zeitz M, Ullrich R, Duchmann R: **Modulation of systemic antigen-specific immune responses by oral antigen in humans.** *Eur J Immunol* 2010, **40**(11):3128–3137.

50. Cassani B, Villablanca EJ, Quintana FJ, Love PE, Lacy-Hulbert A, Blaner WS, Sparwasser T, Snapper SB, Weiner HL, Mora JR: **Gut-tropic T cells that express integrin alpha4beta7 and CCR9 are required for induction of oral immune tolerance in mice.** *Gastroenterology* 2011, **141**(6):2109–2118.

51. Hadis U, Wahl B, Schulz O, Hardtke-Wolenski M, Schippers A, Wagner N, Muller W, Sparwasser T, Forster R, Pabst O: **Intestinal tolerance requires gut homing and expansion of FoxP3+ regulatory T cells in the *lamina propria*.** *Immunity* 2011, **34**(2):237–246.

52. Torgerson TR, Linane A, Moes N, Anover S, Mateo V, Rieux-Laucat F, Hermine O, Vijay S, Gambineri E, Cerf-Bensussan N *et al*: **Severe food allergy as a variant of IPEX syndrome caused by a deletion in a noncoding region of the FOXP3 gene.** *Gastroenterology* 2007, **132**(5):1705–1717.

53. Zennaro D, Scala E, Pomponi D, Caprini E, Arcelli D, Gambineri E, Russo G, Mari A: **Proteomics plus genomics approaches in primary immunodeficiency: the case of immune dysregulation, polyendocrinopathy, enteropathy, X-linked (IPEX) syndrome.** *Clin Exp Immunol* 2012, **167**(1):120–128.

54. Karlsson MR, Rugtveit J, Brandtzaeg P: **Allergen-responsive CD4+CD25+ regulatory T cells in children who have outgrown cow's milk allergy.** *J Exp Med* 2004, **199**(12):1679–1688.

55. Shreffler WG, Wanich N, Moloney M, Nowak-Wegrzyn A, Sampson HA: **Association of allergen-specific regulatory T cells with the onset of clinical tolerance to milk protein.** *J Allergy Clin Immunol* 2009, **123**(1):43–52 e47.

56. Suzuki H, Sekine S, Kataoka K, Pascual DW, Maddaloni M, Kobayashi R, Fujihashi K, Kozono H, McGhee JR: **Ovalbumin-protein sigma 1 M-cell targeting facilitates oral tolerance with reduction of antigen-specific CD4+T cells.** *Gastroenterology* 2008, **135**(3):917–925.

57. Rynda A, Maddaloni M, Mierzejewska D, Ochoa-Reparaz J, Maslanka T, Crist K, Riccardi C, Barszczewska B, Fujihashi K, McGhee JR et al: **Low-dose tolerance is mediated by the microfold cell ligand, reovirus protein sigma1.** J Immunol 2008, **180**(8):5187–5200.

58. Spahn TW, Weiner HL, Rennert PD, Lugering N, Fontana A, Domschke W, Kucharzik T: **Mesenteric lymph nodes are critical for the induction of high-dose oral tolerance in the absence of Peyer's patches.** Eur J Immunol 2002, **32**(4):1109–1113.

59. Spahn TW, Fontana A, Faria AM, Slavin AJ, Eugster HP, Zhang X, Koni PA, Ruddle NH, Flavell RA, Rennert PD et al: **Induction of oral tolerance to cellular immune responses in the absence of Peyer's patches.** Eur J Immunol 2001, **31**(4):1278–1287.

60. Kraus TA, Brimnes J, Muong C, Liu JH, Moran TM, Tappenden KA, Boros P, Mayer L: **Induction of mucosal tolerance in Peyer's patch-deficient, ligated small bowel loops.** J Clin Invest 2005, **115**(8):2234–2243.

61. Macpherson AJ, Smith K: **Mesenteric lymph nodes at the center of immune anatomy.** J Exp Med 2006, **203**(3):497–500.

62. Worbs T, Bode U, Yan S, Hoffmann MW, Hintzen G, Bernhardt G, Forster R, Pabst O: **Oral tolerance originates in the intestinal immune system and relies on antigen carriage by dendritic cells.** J Exp Med 2006, **203**(3):519–527.

63. Viney JL, Mowat AM, O'Malley JM, Williamson E, Fanger NA: **Expanding dendritic cells in vivo enhances the induction of oral tolerance.** J Immunol 1998, **160**(12):5815–5825.

64. Chirdo FG, Millington OR, Beacock-Sharp H, Mowat AM: **Immunomodulatory dendritic cells in intestinal** lamina propria. Eur J Immunol 2005, **35**(6):1831–1840.

65. Mowat AM: **Dendritic cells and immune responses to orally administered antigens.** Vaccine 2005, **23**(15):1797–1799.

66. Peng HJ, Turner MW, Strobel S: **The generation of a 'tolerogen' after the ingestion of ovalbumin is time-dependent and unrelated to serum levels of immunoreactive antigen.** Clin Exp Immunol 1990, **81**(3):510–515.

67. Almqvist N, Lonnqvist A, Hultkrantz S, Rask C, Telemo E: **Serum-derived exosomes from antigen-fed mice prevent allergic sensitization in a model of allergic asthma.** Immunology 2008, **125**(1):21–27.

68. Thomson AW, Knolle PA: **Antigen-presenting cell function in the tolerogenic liver environment.** Nat Rev Immunol 2010, **10**(11):753–766.

69. Menard S, Cerf-Bensussan N, Heyman M: **Multiple facets of intestinal permeability and epithelial handling of dietary antigens.** Mucosal Immunol 2010, **3**(3):247–259.

70. McDole JR, Wheeler LW, McDonald KG, Wang B, Konjufca V, Knoop KA, Newberry RD, Miller MJ: **Goblet cells deliver luminal antigen to CD103+dendritic cells in the small intestine.** Nature 2012, **483**(7389):345–349.

71. Diehl GE, Longman RS, Zhang JX, Breart B, Galan C, Cuesta A, Schwab SR, Littman DR: **Microbiota restricts trafficking of bacteria to mesenteric lymph nodes by CX(3)CR1(hi) cells.** Nature 2013, **494**(7435):116–120.

72. Harden JL, Egilmez NK: **Indoleamine 2,3-dioxygenase and dendritic cell tolerogenicity.** Immunol Invest 2012, **41**(6-7):738–764.

73. Cerovic V, Houston SA, Scott CL, Aumeunier A, Yrlid U, Mowat AM, Milling SW: **Intestinal CD103(-) dendritic cells migrate in lymph and prime effector T cells.** Mucosal Immunol 2013, **6**(1):104–113.

74. Mora JR, Iwata M, Eksteen B, Song SY, Junt T, Senman B, Otipoby KL, Yokota A, Takeuchi H, Ricciardi-Castagnoli P et al: **Generation of gut-homing IgA-secreting B cells by intestinal dendritic cells.** Science 2006, **314**(5802):1157–1160.

75. Hammerschmidt SI, Ahrendt M, Bode U, Wahl B, Kremmer E, Forster R, Pabst O: **Stromal mesenteric lymph node cells are essential for the generation of gut-homing T cells in vivo.** J Exp Med 2008, **205**(11):2483–2490.

76. Hooper LV, Littman DR, Macpherson AJ: **Interactions between the microbiota and the immune system.** Science 2012, **336**(6086):1268–1273.

77. Bashir ME, Louie S, Shi HN, Nagler-Anderson C: **Toll-like receptor 4 signaling by intestinal microbes influences susceptibility to food allergy.** J Immunol 2004, **172**(11):6978–6987.

78. Hill DA, Siracusa MC, Abt MC, Kim BS, Kobuley D, Kubo M, Kambayashi T, Larosa DF, Renner ED, Orange JS et al: **Commensal bacteria-derived signals regulate basophil hematopoiesis and allergic inflammation.** *Nat Med* 2012, **18**(4):538–546.

79. Noval Rivas M, Burton OT, Wise P, Zhang YQ, Hobson SA, Garcia Lloret M, Chehoud C, Kuczynski J, DeSantis T, Warrington J et al: **A microbiota signature associated with experimental food allergy promotes allergic sensitization and anaphylaxis.** *J Allergy Clin Immunol* 2013, **131**(1):201–212.

80. Thompson-Chagoyan OC, Fallani M, Maldonado J, Vieites JM, Khanna S, Edwards C, Dore J, Gil A: **Faecal microbiota and short-chain fatty acid levels in faeces from infants with cow's milk protein allergy.** *Int Arch Allergy Immunol* 2011, **156**(3):325–332.

81. Thompson-Chagoyan OC, Vieites JM, Maldonado J, Edwards C, Gil A: **Changes in faecal microbiota of infants with cow's milk protein allergy-a Spanish prospective case-control 6-month follow-up study.** *Pediatr Allergy Immunol* 2010, **21**(2 Pt 2):e394–400.

82. Rodriguez B, Prioult G, Hacini-Rachinel F, Moine D, Bruttin A, Ngom-Bru C, Labellie C, Nicolis I, Berger B, Mercenier A et al. **Infant gut microbiota is protective against cow's milk allergy in mice despite immature ileal T-cell response.** *FEMS Microbiol Ecol* 2012, **79**(1):192–202.

83. Upadhyay V, Poroyko V, Kim TJ, Devkota S, Fu S, Liu D, Tumanov AV, Koroleva EP, Deng L, Nagler C et al: **Lymphotoxin regulates commensal responses to enable diet-induced obesity.** *Nat Immunol* 2012, **13**(10):947–953.

84. Visness CM, London SJ, Daniels JL, Kaufman JS, Yeatts KB, Siega-Riz AM, Liu AH, Calatroni A, Zeldin DC: **Association of obesity with IgE levels and allergy symptoms in children and adolescents: results from the National Health and Nutrition Examination Survey 2005-2006.** *J Allergy Clin Immunol* 2009, **123**(5):1163–1169, 1169 e1161–1164.

85. Hall JA, Cannons JL, Grainger JR, Dos Santos LM, Hand TW, Naik S, Wohlfert EA, Chou DB, Oldenhove G, Robinson M et al: **Essential role for retinoic acid in the promotion of CD4(+) T cell effector responses via retinoic acid receptor alpha.** *Immunity* 2011, **34**(3):435–447.

86. DePaolo RW, Abadie V, Tang F, Fehlner-Peach H, Hall JA, Wang W, Marietta EV, Kasarda DD, Waldmann TA, Murray JA et al: **Co-adjuvant effects of retinoic acid and IL-15 induce inflammatory immunity to dietary antigens.** *Nature* 2011, **471**(7337):220–224.

87. Yu S, Bruce D, Froicu M, Weaver V, Cantorna MT: **Failure of T cell homing, reduced CD4/CD8alphaalpha intraepithelial lymphocytes, and inflammation in the gut of vitamin D receptor KO mice.** *Proc Natl Acad Sci U S A* 2008, **105**(52):20834–20839.

88. Sharief S, Jariwala S, Kumar J, Muntner P, Melamed ML: **Vitamin D levels and food and environmental allergies in the United States: results from the National Health and Nutrition Examination Survey 2005-2006.** *J Allergy Clin Immunol* 2011, **127**(5):1195–1202.

89. Kiss EA, Vonarbourg C, Kopfmann S, Hobeika E, Finke D, Esser C, Diefenbach A: **Natural aryl hydrocarbon receptor ligands control organogenesis of intestinal lymphoid follicles.** *Science* 2011, **334**(6062):1561–1565.

90. Li Y, Innocentin S, Withers DR, Roberts NA, Gallagher AR, Grigorieva EF, Wilhelm C, Veldhoen M: **Exogenous stimuli maintain intraepithelial lymphocytes via aryl hydrocarbon receptor activation.** *Cell* 2011, **147**(3):629–640.

91. Schulz VJ, Smit JJ, Willemsen KJ, Fiechter D, Hassing I, Bleumink R, Boon L, van den Berg M, van Duursen MB, Pieters RH: **Activation of the aryl hydrocarbon receptor suppresses sensitization in a mouse peanut allergy model.** *Toxicol Sci* 2011, **123**(2):491–500.

92. Schulz VJ, Smit JJ, Huijgen V, Bol-Schoenmakers M, van Roest M, Kruijssen LJ, Fiechter D, Hassing I, Bleumink R, Safe S et al: **Non-dioxin-like AhR ligands in a mouse peanut allergy model.** *Toxicol Sci* 2012, **128**(1):92–102.

93. Quintana FJ, Basso AS, Iglesias AH, Korn T, Farez MF, Bettelli E, Caccamo M, Oukka M, Weiner HL: **Control of T(reg) and T(H)17 cell differentiation by the aryl hydrocarbon receptor.** *Nature* 2008, **453**(7191):65–71.

94. Okada Y, Oh-oka K, Nakamura Y, Ishimaru K, Matsuoka S, Okumura K, Ogawa H, Hisamoto M, Okuda T, Nakao A: **Dietary resveratrol prevents the development of food allergy in mice.** *PLoS One* 2012, **7**(9):e44338.

95. Blazquez AB, Berin MC: **Gastrointestinal dendritic cells promote Th2 skewing via OX40 L**. *J Immunol* 2008, **180**(7):4441–4450.
96. Ziegler SF, Artis D: **Sensing the outside world: TSLP regulates barrier immunity**. *Nat Immunol* 2010, **11**(4):289–293.
97. Chu DK, Llop-Guevara A, Walker TD, Flader K, Goncharova S, Boudreau JE, Moore CL, Seunghyun In T, Waserman S, Coyle AJ *et al*: **IL-33, but not thymic stromal lymphopoietin or IL-25, is central to mite and peanut allergic sensitization**. *J Allergy Clin Immunol* 2013, **131**(1):187–200 e181–188.
98. Yang Z, Sun R, Grinchuk V, Blanco JA, Notari L, Bohl JA, McLean LP, Ramalingam TR, Wynn TA, Urban JF, Jr. *et al*: **IL-33-induced alterations in murine intestinal function and cytokine responses are MyD88, STAT6, and IL-13 dependent**. *Am J Physiol Gastrointest Liver Physiol* 2013, **304**(4):G381–389.

Chapter 2
Intestinal Permeability and Transport of Food Antigens

Contents

Abbreviations

ALA	Alpha-lactalbumin
APC	Antigen-presenting cell
BLG	β-lactoglobulin
CCR	CC chemokine receptor
CXCR	CXC chemokine receptor
DC	Dendritic cell
FcεRII (CD23)	Low affinity Fc receptor II for IgE
FcRn	Neonatal Fc receptor
GALT	Gut-associated lymphoid tissue
GIT	Gastrointestinal tract
IC	Immune complex
IEC	Intestinal epithelial cell
IFN-γ	Interferon gamma
ILF	Isolated lymphoid follicle
Fcgbp	Fc-gamma-binding protein
JAM-A	Junctional adhesion molecule A
LT	Leukotriene
M cells	Microfold cell

T. Ćirković Veličković, M. Gavrović-Jankulović, *Food Allergens,*
Food Microbiology and Food Safety, DOI 10.1007/978-1-4939-0841-7_2,
© Springer Science+Business Media New York 2014

MCP	Monocyte chemoattractant protein
MHC	Major histocompatibility complex
MIIC	Major histocompatibility complex class II-enriched compartment
MLCK	Myosin light chain kinase
MLN	Mesenteric lymph node
MUC	Mucin glycoproteins
OT	Oral tolerance
PG	Prostaglandin
pIgR	Polymeric immunoglobulin receptor
PI3K	Phospho-inositol 3 kinase
PP	Peyer's patch
PRP	Pattern recognition receptor
SIgA	Secretory immunoglobulin A
RELMbeta	Resistin-like molecule beta
TEER	Transepithelial electrical resistance
TFF	Trefoil factor peptide
TGF-β	Transforming growth factor beta
Th	T helper lymphocyte
Tj	Tight junction
TLRs	Toll-like receptors
TNF	Tumour necrosis factor
Treg	T regulatory lymphocyte
Zo	Zonula occludens

Summary The intestinal epithelium forms a selective barrier, which favours fluxes of nutrients, regulates ion and water movements, and limits host contact with the massive intraluminal load of dietary antigens and microbes. However, this barrier is not fully impermeable to macromolecules; in the steady state, the transepithelial passage of small amounts of food-derived antigens and microorganisms participates in the induction of a homeostatic immune response dominated by an immune tolerance to dietary antigens and the local production of secretory immunoglobulin A. Conversely, primary or secondary defects of the intestinal barrier can lead to excessive entrance of dietary or microbe-derived macromolecules, which are putative contributors to the pathogenesis of a wide range of human diseases, including food allergy and inflammatory bowel disease, and could even be related to autoimmune diseases and metabolic syndrome. Although a majority of dietary proteins are totally degraded by digestive enzymes and are absorbed in the form of nutrients (amino acids or dipeptides/tripeptides), some can resist both the low pH of the gastric fluid and proteolysis by enzymes, providing large immunogenic peptides or even intact proteins to reach the small intestinal lumen.

2.1 Intestinal Permeability of Antigens

The intestinal epithelium forms a selective barrier, which favours fluxes of nutrients, regulates ion and water movements, and limits host contact with the massive intraluminal load of dietary antigens and microbes. However, this barrier is not fully impermeable to macromolecules; in the steady state, the transepithelial passage of small amounts of food-derived antigens and microorganisms participates in the induction of a homeostatic immune response dominated by immune tolerance to dietary antigens and the local production of secretory immunoglobulin A (SIgA). Conversely, primary or secondary defects of the intestinal barrier can lead to excessive entrance of dietary or microbe-derived macromolecules, which are putative contributors to the pathogenesis of a wide range of human diseases, including food allergy and inflammatory bowel disease, and could even be related to autoimmune diseases and metabolic syndrome [1, 2]. The intestinal barrier and more particularly the paracellular pathway have recently been reinforced and suggested as a therapeutic target to treat or prevent diseases driven by luminal antigens [1].

Dietary antigens are available for intestinal transport: Although the majority of dietary proteins are totally degraded by digestive enzymes and are absorbed in the form of nutrients (amino acids or dipeptides/tripeptides), some can resist both the low pH of the gastric fluid and proteolysis by enzymes, providing large immunogenic peptides or even intact proteins to reach the small intestinal lumen (Fig. 2.1).

The intestinal transport of molecules from the intestinal lumen to *lamina propria* can occur through two distinct mechanisms: the paracellular diffusion through tight junctions (TJs) between adjacent intestinal epithelial cells (IECs) and the transcellular transport involving endocytosis/exocytosis (transcytosis) mediated or not by membrane receptors.

2.1.1 Paracellular Transport Pathways

There are five types of gut epithelial cells: absorptive columnar cells (enterocytes), goblet, endocrine, Paneth, and M (microfold) cells. Epithelial cohesion and polarity are maintained by the apical junctional complex, which is composed of tight and adherens junctions, and by the sub-adjacent desmosomes (Fig. 2.2) [1]. TJs are located apically and represent the barrier and the rate-limiting factor for the paracellular permeation of molecules. Paracellular transport across the epithelial barrier is highly regulated and is able to leave access not only to small molecules but also to larger molecules, including small peptides and bacterial lipopolysaccharides through the 'leak pathway' [4]. The width of the lateral space below the TJ is around 75 Å, being wider in comparison to 4–9 Å for the TJ pores in villi or 50–60 Å for the TJ pores in crypts [5]. The other structures shown in Fig. 2.2 maintain cell proximity, but are not rate limiting as TJs. The difference in the width of TJ pores in villi and crypts also suggests a decreasing gradient of paracellular permeability from crypt to villous IECs.

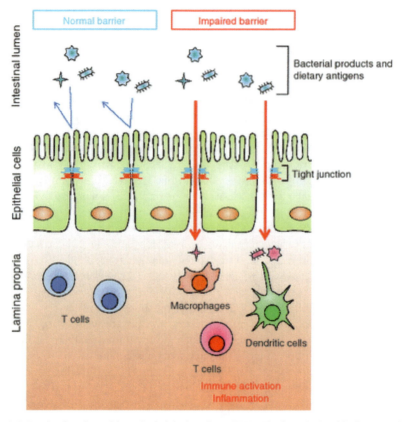

Fig. 2.1 Barrier function of intestinal tight junctions (*TJs*). The intestinal epithelium provides a physical barrier to luminal bacteria, toxins, and antigens. The barrier is organized by different barrier components, including the *TJs*. The *TJs* regulate the paracellular passage of ions, solutes, and water between adjacent cells. Luminal noxious macromolecules cannot penetrate the epithelium because of the *TJ* barrier; however, *TJ* barrier impairment allows the passage of noxious molecules, which can induce the excessive activation of mucosal immune cells and inflammation. Therefore, intestinal barrier defects are associated with the initiation and development of various intestinal and systemic diseases. (The figure is reprinted from Ref [3] with kind permission from Springer Science and Business Media)

TJs are multi-protein complexes mainly composed of proteins, such as membrane proteins occludin and claudins, junctional adhesion molecule A (JAM-A), and tricellulin, which control the permeability of TJs.

Occludin interacts directly with membrane-associated *zonula occludens* (ZO-1, ZO-2, and ZO-3) proteins that regulate the perijunctional actinomyosin ring. Claudins determine the charge selectivity of the paracellular pathway [6]. Claudin-2 forms cation-selective channels in the intestinal TJs. A tightening function has been described for claudin-1, -3, -4, -5, -8, -11, -14, and -19 using knockout mouse models or a range of assays. Other claudins, such as claudin-2, -7, -10, -15, and -16, are pore-forming molecules likely to increase paracellular permeability [6]. The junctional protein JAM-A is also involved in the control of mucosal homeostasis by regulating the integrity and

Lumen

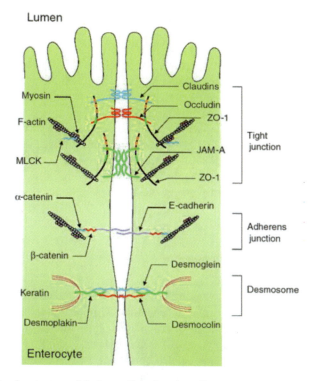

Fig. 2.2 Molecular structure of the intercellular junction of intestinal epithelial cells. The intercellular junctions of intestinal epithelial cells are sealed by different protein complexes, including tight junctions (*TJs*), adherens junctions, and desmosomes. The *TJs*, multiple protein complexes, are located at the apical ends of the lateral membranes of intestinal epithelial cells. The *TJ* complex consists of transmembrane and intracellular scaffold proteins. The extracellular loops of the transmembrane proteins (occludin, claudins, junctional adhesion molecules (*JAMs*), and tricellulin) create a permselective barrier in the paracellular pathways by haemophilic and heterophilic interactions with adjacent cells. The intracellular domains of the transmembrane proteins interact with the intracellular scaffold proteins such as zonula occludens (*ZO*) proteins and cingulin, which in turn anchor the transmembrane proteins to the actin cytoskeleton. Myosin light chain kinase (*MLCK*) is associated with the perijunctional actinomyosin rings and regulates paracellular permeability through myosin contractility. The adherens juntions along with desmosomes provide strong adhesive bonds between the epithelial cells and also intercellular communication, but do not determine paracellular permeability. (The figure is reproduced from Ref [3] with kind permission from Springer Science and Business Media)

permeability of epithelial barrier function and leukocyte migration [7]. A decreased expression of JAM-A and an increased inflammatory cytokine production have been reported in inflammatory bowel disease [8, 9]. In JAM-A-deficient mice, increased intestinal permeability is related to the upregulation of claudin-10 and claudin-15 [7].

Tricellulin has a critical role in the formation of the epithelial barrier [10]. It is an integral membrane protein contributing to the structure and function of tricellular contacts between neighbouring cells. Sodium caprate's enhancing effect on intestinal drug uptake is based on increased permeability in tricellular cell contacts, mediated by reversible removal of tricellulin from the tricellular TJ [11].

Paracellular permeability is related to the pores in the epithelial TJ determining a high-capacity, size-restricted pathway and a low-capacity, size-independent (or less restrictive) pathway that might be due to fixed (e.g. tricellular junctions, larger pores) or transient breaks (e.g. apoptosis) in the epithelial monolayer. The paracellular pathway accepts molecules with molecular mass < 600 Da (Fig. 2.1); permeation activity through this pathway can be measured from the diffusion of small inert probes (mannitol = 6.7 Å, lactulose = 9.5 Å). A large number of small pores together with a small number of large pores could explain the higher permeation of small-sized markers. Recent findings suggest that both small and large pores are defined by different TJ proteins, such as claudins and tricellulin, respectively. The paracellular diffusion of small molecules through TJ pores is driven by water movement due to transepithelial electrochemical or osmotic gradients.

2.1.2   Transcellular Transport Pathways

A transcellular transport pathway allows large antigenic molecules to gain access to the subepithelial compartment and to interact with local immune cells. Macromolecules can be sampled from the intestinal lumen into enterocytes by a vesicular transport (fluid phase or receptor-mediated endocytosis) and released basolaterally. M cells located in the follicle-associated epithelium of Peyer's patches (PPs) are involved in the transcytosis of bacteria. M cells express receptors on apical surfaces (Toll-like receptors, lectin-like microbial adhesins, $\alpha5\beta1$ integrin, platelet-activating factor receptor, or glycoprotein-2) thus being specialized in transcytosis of bacteria. Columnar enterocytes, the major cell type in the small intestinal epithelium, can sample, transport, and/or process soluble antigens all along the intestine.

It has been shown that antigen transport through afferent lymphatics into the draining mesenteric lymph nodes (MLNs) is obligatory for oral tolerance (OT) induction. PPs are not mandatory for the development of OT. A rapid antigen uptake by dendritic cells (DCs) in the small intestinal *lamina propria* occurs after feeding mice with ovalbumin [2].

PPs seem to be dispensable for the induction of tolerance against soluble antigens: Oral administration of a soluble antigen in PP-deficient mice results in the same frequency of DCs and macrophages in MLNs, peripheral lymph nodes, and spleen [12]; knockout mice devoid of PPs, but with fully developed MLNs, are competent in establishing systemic tolerance against an antigen given by the oral route [13]. The data show that in the presence of MLNs, PPs are not required for OT induction and that the presence of MLNs is sufficient for OT induction [13]. Taken collectively, this suggests that antigen-presenting cells (APCs) loaded with antigens emanating from isolated lymphoid follicles (ILF; or from the mucosa) are sufficient to induce OT.

The transcellular transport of large particles (Fig. 2.3) has been traditionally ascribed to M cells overlying PPs and ILF in the distal part of the intestine [1]. Intestinal DCs have the capacity to sample bacteria directly in the intestinal lumen by

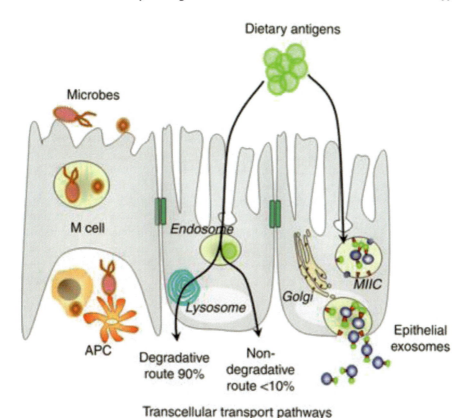

Fig. 2.3 Transcellular transport pathways. Under steady-state condition, molecules of molecular weight (MW) >600 Da (such as food antigens, peptides) are sampled by the epithelial cells by endocytosis at the apical membrane and transcytosis toward the lamina propria. During transcytosis, full-length peptides or proteins are partly degraded in acidic and lysosomal compartments and released in the form of amino acids (total degradation) or breakdown products (partial degradation) at the basolateral pole of enterocytes. Early *endosomes* containing partially degraded food antigens meet the major histocompatibility complex (MHC) class II-enriched compartment (*MIIC*) where exogenous peptides are loaded on MHC class II molecules. Inward invagination of *MIIC* compartment lead to the formation of exosomes, which are small membrane vesicles (40–90 nm) bearing MHC class II/peptide complexes at their surface. Exosomes can diffuse in the basement membrane and interact with local immune cells. Exosome-bound peptides are much more potent than free peptides to interact with dendritic cells and stimulate peptide presentation to T cells. (Figure reprinted by permission from Macmillan Publishers Ltd from Ref [1])

extending dendrites between epithelial cells. The role of M cells in the sampling of soluble antigens is not exclusive, and food antigens present in the proximal intestine can be transported through columnar enterocytes. In ex vivo studies of animal and human intestinal mucosa, the uptake of food antigens by IECs was associated with a powerful epithelial degradation.

Using chambers with tritiated food proteins (^{3}H-lysine) added to the apical compartment, transport and degradation during transcytosis could be quantified by analysis of ^{3}H-labeled fragments released in the basolateral compartment. These studies indicated that only small amounts of intact protein are transcytosed, around 0.1% of luminal concentration. Epithelial cells process proteins into peptides of ~1.500-Da molecular weight, which is a size compatible with the peptide-binding pocket of antigen-presenting major histocompatibility complex (MHC) class II molecules. Large proteins taken up by IECs were released on their basolateral side either as immunogenic peptides (~40%) or fully degraded into amino acids (~50%), with only a minor fraction crossing the epithelium in their intact form (<10%). Large peptides or intact proteins released into the *lamina propria* might then be taken up by local APCs. It is also possible that protection from the total degradation might occur during the transcytosis process.

Antigenic material may also reach the *lamina propria* within exosomes, small membrane vesicles (~80 nm in diameter), derived from MHC class II-expressing enterocytes. Exosomes are formed when MHC class II-loading compartments fuse with endosomes containing only partially degraded proteins (Fig. 2.3). It has been shown that exosomes can be taken up efficiently by APCs. It is uncertain whether this takes place in vivo, and the relative importance of this mechanism in the induction or the maintenance of OT compared with the other APCs remains difficult to evaluate.

In professional APCs, exogenous antigens are endocytosed at the cell surface and are processed in early endosomes before reaching MHC class II-enriched compartments (MIICs) where antigen-derived peptides are loaded on MHC class II molecules. In this compartment, exosomes are formed by inward invagination of the MIIC-limiting membrane and they carry MHC class II/peptide complexes at their surface [14]. MIICs can either be directed to the lysosomal compartment or fuse with the plasma membrane and release exosomes into the extracellular medium.

Epithelium-derived exosomes may be potent vehicles involved in the intestinal antigen presentation. In vitro studies have shown that peptides released when bound to exosomes interacted very efficiently with DCs and could potently stimulate antigen-specific T cell clones at much lower concentration than free peptides [15]. In vivo administration of epithelial exosomes can prime an immunogenic or a tolerogenic immunogenic response [14]. Recent studies demonstrated that the serum or isolated serum exosomes obtained from ovalbumin-fed mice and administered intraperitoneally to naive recipient mice abrogated allergic sensitization in the recipients [16]. It also seems that in vivo outcome of immune stimulation by epithelium-derived exosomes depends on the nature of the APCs, and on their conditioning by epithelium-derived factors [17]. In the intestine, both the mucosal microenvironment and local effector cells are probably key players in determining the outcome of the immune response to exosome-derived epitopes.

The transcytosis of food antigens occurs primarily by a fluid-phase endocytosis of proteins at the apical membrane of enterocytes; however, under different circumstances, pathogenic antigens can access the mucosa through the expression of immunoglobulin (Ig) receptors at the apical surface of enterocytes, thereby allowing their entry in the form of immune complexes (ICs).

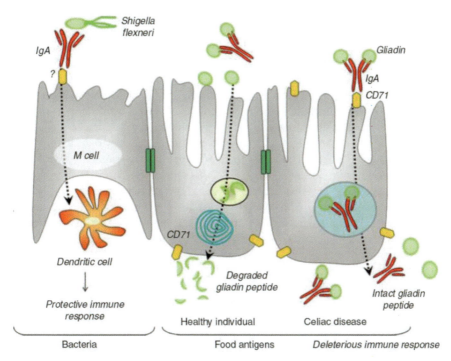

Fig. 2.4 Immunoglobulin (*Ig*)*A*-mediated retrotransport of luminal antigens. *IgA* is a protective mucosal immunoglobulin secreted in the intestinal lumen through polymeric immunoglobulin receptor (pIgR) in the form of secretory IgA (SIgA). Whereas the major role of SIgA is to contain microbial and food antigens in the intestinal lumen, in some pathological situations, an abnormal retrotransport of SIgA immune complexes can allow bacterial or food antigens entry in the intestinal mucosa, with various outcomes. Indeed, SIgA can mediate the intestinal entry of SIgA/*Shigella flexneri* immune complexes through M cells and interact with dendritic cells, inducing an inflammatory response aimed at improving bacterial clearance and the restoration of intestinal homeostasis. In celiac disease, however, SIgA allows the protected transcytosis of gliadin peptides, a mechanism more likely to trigger exacerbated adaptive and innate immune responses in view of the constant flow of gluten in the gut and to precipitate mucosal lesions. Indeed, whereas in healthy individuals, undigested gliadin peptides are taken up by non-specific endocytosis in enterocytes and entirely degraded/detoxified during transepithelial transport, in contrast, in active celiac disease, the ectopic expression of *CD71* (the transferrin receptor also known as *IgA* receptor) at the apical membrane of epithelial cells favours the retrotransport of IgA immune complexes and inappropriate immune responses. (Figure reprinted by permission from Macmillan Publishers Ltd from Ref [1])

2.1.3   Antigen Transcytosis Through Immune Complexes

IgA-Mediated Transport   IgA is the most representative Ig isotype at the mucosal surface. A common receptor-mediated IgA transport mechanism in the intestine is the basal-to-apical secretion of dimeric IgA in the form of secretory SIgA through the polymeric Ig receptor [18]. The major role of SIgA is to bind and keep microbial and food antigens in the intestinal lumen (Fig. 2.4). Although SIgAs are mainly devoted to restricting the entry of exogenous antigens in the intestinal mucosa, some cases of apical-to-basal retrotransport have been reported. SIgA also triggers migration of DCs to the T cell-rich regions of PPs, and regulates expression of CD80 and CD86 on DCs in PPs, MLNs, and spleen. These results provide evidence that mucosal SIgA re-entering the body exerts a function of Ag delivery that contributes to effector and/or regulatory pathways characteristic of the intestinal mucosal compartment [19]. In some pathological situations, an abnormal retrotransport of SIgA ICs can allow bacterial or food antigens entry into the intestinal mucosa. This retrotranscytosis of SIgA–gliadin complexes may promote the entry of harmful gliadin peptides into the intestinal mucosa, thereby triggering an immune response and perpetuating intestinal inflammation [20]. This mechanism is likely to trigger an exacerbated immune response due to a constant flow of gluten in the gut. In healthy individuals, undigested gliadin peptides are taken up by non-specific endocytosis in enterocytes and degraded during transepithelial transport. However, in active celiac disease, the expression of CD71 (the transferrin receptor, or IgA receptor) at the apical membrane of epithelial cells favours the retrotransport of IgA ICs and inappropriate immune responses. The presence of large aggregates of gliadin-specific IgA in duodenal secretions, *lamina propria,* and serum of celiac patients could provide a danger signal promoting the rupture of OT and/or triggering tissue damage [20, 21].

IgE-Mediated Allergen Transport In food allergy, the low-affinity receptor for IgE, (FcεRII, CD23), is abnormally overexpressed in IECs [22]. Allergens complexed with IgE can bypass epithelial lysosomal degradation, resulting in the entry of a large amount of intact allergens into the mucosa (Fig. 2.5) [23]. High levels of interleukin (IL)-4, a Th2-type cytokine elevated in allergic diseases, upregulates the expression of CD23 at the apical side of IECs. An overexpression of CD23 at the apical side of enterocytes can drive the transport of intact IgE/allergen ICs from the intestinal lumen to the *lamina propria.* Evidence shows that IgE/CD23-mediated transport of allergens protect them from degradation in sensitized animals [24]. Large amounts of transported allergens into *lamina propria* may trigger mast cell degranulation and induce an allergic inflammatory cascade. This mechanism could participate in the rapid onset of intestinal symptoms in IgE-dependent food allergy [25].

IgG-Mediated Transport of Antigens IgGs are not classical secretory antibodies. Gastrointestinal secretions contain significant amounts of IgG suggesting their protective role [26]. Now, it is widely accepted that presence of IgG in intestinal lumen is not related to the non-specific diffusion of IgG through TJ as often suggested, but to a specific mechanism.

IgGs have initially been shown to bind the neonatal Fc receptor on IECs (FcRn) [27] in the acidic environment close to the apical membrane or in early endosomes of enterocytes (Fig. 2.5) [1]. This receptor-mediated transcytosis allows a protected transport of IgG and their release on the basal side of enterocytes where neutral pH induces their dissociation from the receptor [27]. In vitro studies have indicated that IgG ICs can also be shuttled from the apical to the basal pole of enterocytes through FcRn and vice versa [1].

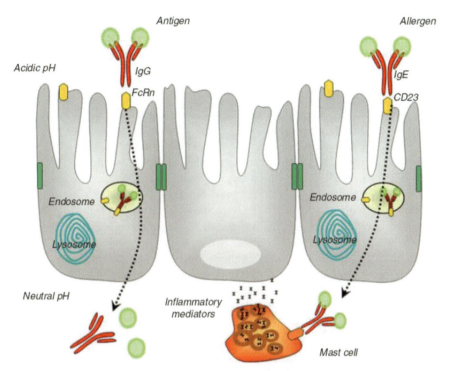

Fig. 2.5 *IgE*- and *IgG*-mediated transport of antigens. Immunoglobulin (*Ig*)*G*-mediated transport of antigens. Although *IgGs* are not classical secretory antibodies, their presence in the intestinal lumen suggests a protective role. *IgGs* have initially been shown to bind the neonatal Fc receptor on intestinal epithelial cells (*FcRn*) in the acidic environment close to the apical membrane or in early endosomes of enterocytes. This receptor-mediated transcytosis allows a protected transport of *IgG* and their release on the basal side of enterocytes where neutral pH induces their dissociation from the receptor. In vitro studies have indicated that IgG immune complexes can also be shuttled from the apical to the basal pole of enterocytes through FcRn and vice versa, although the incidence of *IgG* immune complexes in terms of immune response is not clearly established. *IgE*-mediated allergen transport. In food allergy, the low-affinity receptor for *IgE*, *CD23*, is abnormally overexpressed in intestinal epithelial cells, in humans and murine models of allergy. An overexpression of *CD23* at the apical side of enterocytes can drive the transport of intact *IgE*/allergen immune complexes from the intestinal lumen to the lamina propria, a phenomenon triggering mast cell degranulation and allergic inflammatory cascade. (Figure reprinted by permission from Macmillan Publishers Ltd from Ref [1])

2.1.4 Transport of Antigens via CX3CR1hi Cells

A particular subset of myeloid cells CD103-CX3CRhi1 is able to probe the intestinal lumen by extending their dendrites across the epithelial barrier without perturbing TJs and epithelial integrity. The transepithelial dendrites from CX-3CR1hi myeloid cells in the *lamina propria* have been shown to interact with bacteria in the lumen, and they may play a role in inducing tolerance to commensal bacteria. It has been speculated that these CX3CR1hi cells might facilitate antigen uptake in vivo, but the frequency of transepithelial protrusions varies markedly between mouse strains and between the various segments of the intestine. These cells may play an important role in host protection during pathogen infection, but do not appear to play a role in antigen capture by *lamina propria* DCs in the steady state [28]. These cells do not migrate from *lamina propria* to MLNs and cannot present luminal antigen to naïve T cells. However, their auxiliary role in passing the antigen to neighbouring migratory DCs for transport and presentation cannot be excluded.

The intestinal immune system is instructed by the microbiota to limit responses to luminal antigens. It has been demonstrated that, at steady state, the microbiota inhibits the transport of both commensal and pathogenic bacteria from the lumen to the MLNs. However, in the absence of conditions of antibiotic-induced dysbiosis, non-invasive bacteria were trafficked to the MLNs in a CC chemokine receptor 7 (CCR7)-dependent manner, and induced both T cell responses and IgA production. Trafficking was carried out by CX3CR1(int) mononuclear phagocytes, an intestinal cell population previously reported to be non-migratory. These findings define a central role for commensals in regulating the migration to the MLNs of CX(3)CR1(int) mononuclear phagocytes endowed with the ability to capture luminal bacteria, thereby compartmentalizing the intestinal immune response to avoid inflammation [29].

2.1.5   Transport of Antigens via Goblet Cells

Recent discovery of goblet cells as a major conduit of antigens from the lumen to the CD103 + DCs shed light on the process of antigen sampling and OT induction. CD103 + cells have regulatory feature and can migrate to the MLNs and prime induction of regulatory T cells. Thus, it has been proposed that antigen delivered by goblet cell to intestinal DCs with tolerogenic phenotype contributes to intestinal immune system homeostasis [28]. Due to the lack of CX3CR1, these cells cannot make transepithelial protrusions and depend on the auxiliary pathways for interaction with an antigen. It seems that goblet cells preferentially deliver its cargo to the CD103 + DCs. In vivo data showed that luminal antigen was captured preferentially by CD103 + DCs at a proportion of roughly 10:1 over CD103− DCs and rarely co-localized with plasmacytoid DCs [30].

2.1.6   Role of Mucin

The mucus layer coating the gastrointestinal tract is the front line of innate host defence, being rich in secretory products of intestinal goblet cells. Goblet cells synthesize secretory mucin glycoproteins (MUC2) and bioactive molecules such as resistin-like molecule beta (RELMbeta), epithelial membrane-bound mucins (MUC1, MUC3, and MUC17), trefoil factor peptides (TFF), and Fc-gamma-binding protein

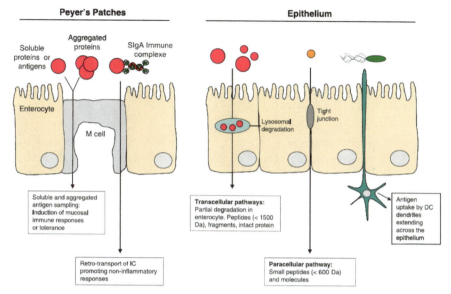

Fig. 2.6 Differential pathways of antigen sampling in the healthy *epithelium*. Organization of the gut epithelium makes it an efficient tight barrier with filtering properties against the entry of pathogenic agents and possibly harmful molecules such as toxins and allergens. Microfold (M) cells present on the surface of the follicle-associated epithelium in intestinal *Peyer's patches* transport particulate antigens and aggregated proteins for presentation by local dendritic cells, resulting in the onset of a tolerogenic type of immune responses under steady-state conditions. IgA-based immune complexes are similarly taken up by M cells and promote the induction of non-inflammatory cytokines (TGF-β and IL-10), ensuring low reactivity against the transported antigen. Partially degraded proteins and a small proportion of intact proteins are taken up by enterocytes. Degradation along the phago-lysosomal pathway occurs, thus resulting in the loss of potentially allergenic properties. Paracellular selective leakage provides access to ions, amino acids, and carbohydrates, which are important in ensuring liquid fluxes and maintenance of transepithelial gradients. Direct intestinal sampling of bacterial antigens by dendritic cells extending their dendrites across the tight epithelium and release of exosome vesicles (*not drawn*) represent other plausible pathways. *TGF-β* transforming growth factor beta, *SigA* secretory IgA. (Figure is reprinted from Ref [2] with permission from John Wiley and Sons)

(Fcgbp). The MUC2 mucin protein forms trimers by disulphide bonding coupled with cross-linking provided by TFF and Fcgbp proteins, resulting in a highly viscous extracellular layer. Colonization by commensal intestinal microbiota is limited to an outer mucus layer. Defective mucus layers resulting from the lack of MUC2 mucin, or from deletion of core mucin glycosyltransferase enzymes in mice, result in increased bacterial adhesion to the surface epithelium and increased intestinal permeability [31].

Thus, orally administered inert particles or non-pathogenic bacteria localize preferentially in organized tissues of the gut-associated lymphoid tissue (GALT) rather than the villous *lamina propria,* indicating that the efficient uptake of particulate material usually requires follicle-associated epithelia and M cells. In conclusion, the nature of the antigen determines its route of uptake (Fig. 2.6). Particulate material and microbiota mostly enter into GALT by M-cell-mediated transcytosis, whereas soluble antigens induce the OT after being taken up by DCs in *lamina propria* and, to a lesser extent, the GALT.

2.2   Protein Degradation in the Gastrointestinal Tract

Dietary proteins, including food allergens, are digested in the gastrointestinal tract [32–37]. Further degradation of food allergens occurs during the passage through enterocytes. Thereby, protein digestion, but also the uptake and degradation process of proteins, is crucial for the induction of a pathological immune response, as well as induction of tolerogenic mechanisms.

The process of protein digestion starts in the mouth. Although lacking proteolytic enzymes, this process helps further digestion of proteins by breaking the food structure and macerating food components in saliva. The main proteolysis of proteins occurs in the gut with the help of pepsin and hydrochloric acid, which breaks the three-dimensional structure of proteins, induces protein unfolding, and helps pepsin digest the protein at the susceptible sites. Further protein digestion continues with the help of pancreatic proteases, trypsin, chymotrypsin, and carboxypeptidase A. Most of the dietary proteins are degraded in the gastrointestinal tract into short oligopeptides and single amino acids. Certain proteins, mostly due to their stable and compact fold, are resistant to the action of gut acid and proteases. From the early work of Astwood et al. [38], resistance of allergens to digestion, especially to pepsin action, was proposed to be related to allergenic potential of dietary proteins.

2.3   Antigen Presentation in the Gastrointestinal System: Role of APCs of the Gut

The intestinal immune system responds to ingested antigens in a variety of ways, ranging from tolerance to full immunity. How T cells are instructed to make these differential responses is still unclear. DCs sample enteric antigens in the *lamina propria* and PPs, and transport them within the patch or to mesenteric nodes where they are presented to lymphocytes. It is probable that DCs also transmit information that influences the outcome of T cell activation, but the nature of this information and the factors in the intestine that regulate DCs behaviour and properties are far from clear.

Professional APCs, such as activated DCs, activated macrophages, and activated B cells express co-stimulatory molecules and, therefore, can not only prime but also activate T cells. On the other hand, APCs without co-stimulatory potential (naïve B cells, resting DCs, and MHC II-expressing tissue cells) induce tolerance, which is mediated by the suppression of T cell effector and helper function, also referred to as anergy or peripheral tolerance. Macrophages, DCs, epithelial cells, and mast cells also express pattern recognition receptors (PRPs) on their surface, i.e., Toll-like receptors (TLRs) that can recognize allergens and secrete various cytokines, including IL-13 [39].

The gastrointestinal system plays a central role in immune system homeostasis. GALT represents almost 70% of the entire immune system.

The gastrointestinal mucosal immune system has a demanding task in distinguishing between desired and harmless dietary antigens and intestinal commensal bacterial populations and, on the other hand, possible harmful molecules and pathogenic organisms. The specific environment of the intestine lumen leads to the development of a powerful tolerogenic mechanism.

A very important portal for presenting antigens in the gut lumen are organized lymphoid tissue structures known as PPs, which are distributed in subepithelia along the intestinal wall, below specialized M cells. Antigens entering through the M-cell-dependent mechanism are eventually delivered to underlying APCs, DCs, or macrophages, which migrate into the PPs, priming immune response that way.

Another important portal for antigen presentation is via subepithelial DCs. It was found that a much higher proportion of orally administered soluble proteins associates with DCs in the *lamina propria* than in the PPs [40]. With respect to invasive or non-invasive pathogens, DCs are able to induce tolerance, or reaction [41, 42].

Antigens can be taken up and presented to the immune system by IECs, which might act as non-professional APCs and modulate local immune response through the activation of intraepithelial CD8(+) T cells with regulatory functions [43].

Epithelial cells process proteins into peptides of ~1,500-Da molecular weight, which is a size compatible with the peptide-binding pocket of antigen-presenting MHC class II molecules. Large proteins taken up by IECs were released on their basolateral side either as immunogenic peptides or fully degraded into amino acids, with only a minor fraction crossing the epithelium in their intact form (<10%). Large peptides or intact proteins released into the *lamina propria* might then be taken up by local APCs. It has also been shown that transcytosis process in the form of ICs provides protection from the total degradation of an antigen. Allergens complexed with IgE can bypass the epithelial lysosomal degradation, resulting in the entry of a large amount of intact allergens into the mucosa.

Hyperpermeability of the gastrointestinal mucosal barrier results in the enhanced transport of intact and degraded antigens across the gastrointestinal mucosal barrier, which could favour food protein sensitisation and food allergy in susceptible individuals [44].

2.4 Intestinal Permeability in Food-Allergic Individuals

Numerous intestinal diseases are characterized by the immune cell activation and compromised epithelial barrier function. Increase in intestinal permeability has been implicated in the pathogenesis of celiac disease, Crohn's disease, type I diabetes, and food allergy [45–48].

The sugar permeation test is a convenient test to measure changes of intestinal permeability to small molecules. Urinary excretion of non-metabolizable markers, disaccharides and monosaccharides (lactulose and mannitol), and ratio of their excretion is a basis for measurement of intestinal permeability. Lactulose and mannitol ratio is the most commonly used test for assessment of small intestinal permeability, and the most reliable method for measurement of lactulose and mannitol concentrations in the urine is high-performance liquid chromatography [45]. After the measurement of concentration of probes in the urine, the results are expressed as the ratio of percentage excretion of the ingested dose of lactulose and mannitol in the urine (L/M ratio). Intestinal permeability is evaluated by measuring the urinary excretion of orally absorbed (0.1 g/kg of weight) mannitol and lactulose. Mannitol is considered to be a marker of absorption of small molecules, while lactulose is a marker of abnormal absorption of large molecules.

Early studies of food-allergic subjects demonstrated that intestinal permeability towards sugars increased in patients after an oral challenge or in patients before a strict exclusion diet. In 12 children with cow's milk-sensitive enteropathy under an exclusion diet, the L/M ratio was comparable with that of controls during fasting and exhibited a threefold rise during a provocation intestinal permeability test with milk [49]. The value of sugar permeation test returned to normal after a strict exclusion diet, suggesting that an increased intestinal permeability might also constitute a cause of allergic reactions [49]. Another study demonstrated increased sugar permeability values in food-allergic patients on an exclusion diet for 6 months thus arguing that the allergic status of the intestinal mucosa remains present even long after the last allergen exposure [50, 51]. Ventura et al. demonstrated that impaired intestinal permeability was present in all subjects with adverse reactions to food. In addition, for the first time, they reported a statistically significant association between the severity of referred clinical symptoms and the increasing L/M ratio [51]. The intestinal permeability test for the diagnosis of food allergies seems to be a sensitive and non-invasive test that is well suited to the paediatric practice [52].

However, the sugar permeation test reflects the passage of a small, uncharged molecule through the mucosa and it does not imply a high permeability to larger molecules, such as protein antigens. In particular, transcellular passage of protein antigens has often been studied in in vitro models of mucosal epithelium. Tumour-derived cell lines, such as Caco-2, T84, and HT-29, are widely used despite many drawbacks. The most common model system is Caco-2 cell monolayer [53]. Caco-2 cells differentiate and adopt many features of epithelial cells, such as TJs and apical and basal cell polarity. The functional integrity of monolayers grown on a filter in a transwell plate is determined by transepithelial electrical resistance (TEER) measurements.

However, the Caco-2 monolayer does not mimic the epithelium with regard to mucus secretion. Thereby, different cell lines, particularly HT-29, were employed in transepithelial studies mimicking presence of mucus [53]. In addition, co-cultures of Caco-2 cells and mucus-secreting tumour cell lines have been developed as sophisticated model systems of bioavailability and transport studies. Ex vivo systems use a chamber and a section of intestinal tissue to study permeation of both small molecules and protein antigens. That chamber consists of a cylindrical tissue holder to which the electrodes and plumbing attach. Tissues are mounted directly onto it, and are compressed between the two chamber halves. The enclosed baths (apical, basolateral) are perfused via a glass circulation reservoir that mounts above the chamber. Studies of protein passage through the jejunal biopsies demonstrated that the transcellular passage of proteins is increased in allergic children and return back to normal after the exclusion diet.

Animal models, especially rodents, have also often been used in in vivo studies of the protein permeability through the epithelial barrier. The probe molecule often used in in vivo studies of intestinal permeability is polyethylene glycol. Polyethylene glycol was used to investigate intestinal absorption in patients with eczema and evidence of food allergy and patients with eczema alone [54]. In both groups, absorption of large molecules was greater than in the normal subjects. There was no difference in the absorption between eczema patients with or without food allergy. These results suggested that there was an intestinal mucosal defect in eczema which exists whether or not there is a coexistent food allergy.

Taken collectively, these data indicate that intestinal permeability towards small molecules and protein antigens is increased during the effector phase of allergic reactions (Fig. 2.7).

Studies conducted in rodent models of food allergies demonstrated that the transcellular pathway of allergen occurs during the sensitization phase of allergy, while during the effector phase, allergen takes the paracellular route of passage. This route, otherwise sealed by TJs, depends on the presence of mast cells and chymase secretion from activated mast cells. Presence of chymase within 20–30 min of allergen exposure allows intact allergens to cross the disturbed barrier. A mast cell granule chymase increases epithelial permeability via a paracellular route and implies that the substrate may be a protein, or proteins, in the epithelial junctional complex [55, 56]. It has also been shown that mast cell-specific protease affects the TJs' integrity by degrading its integral component, occludin [57]. In addition, release of histamine affects physiology of the mucosa by promoting secretion of mucus and electrolytes.

Mast cells upon activation secrete a wide range of mediators of intestinal epithelium permeability, most important being cytokines interleukin-13 (IL-13), IL-8, and tumour necrosis factor-alpha (TNF-alpha). TNF-alpha and IL-13 induce barrier defects that are associated with myosin light chain kinase (MLCK) activation [58] and increased claudin-2 expression in cultured intestinal epithelial monolayers [59]. MLCK activation alters size selectivity to enhance paracellular flux of uncharged macromolecules. In contrast, IL-13-dependent claudin-2 expression increases paracellular cation flux without altering TJ size selectivity, and it is unaffected by MLCK inhibition. In vivo, MLCK activation increases paracellular flux of uncharged macromolecules and also triggers IL-13 expression, claudin-2 synthesis, and increases paracellular cation flux. Reversible, MLCK-dependent permeability increases mucosal immune activation that, in turn, affects TJ to establish long-lasting barrier defects.

IL-4 and IL-13 exhibit a functional overlap that can be explained by the sharing of the receptor component interleukin-4 receptor (IL-4R) alpha. Binding of IL-4 to

Phase 1: Increased transcellular permeability
in sensitized epithelium

Phase 2: Increased paracellular permeability after
degranulation of mast cells

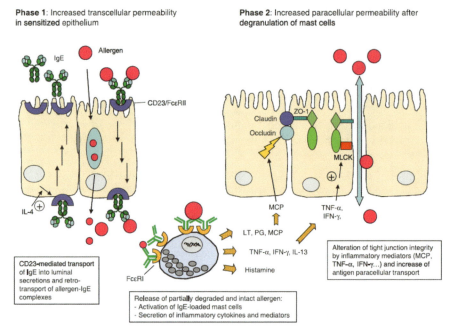

Fig. 2.7 Mechanism of increased intestinal permeability in the sensitized epithelium. In allergic subjects, mast cells loaded with allergen-specific *IgE* are present in the lamina propria. Small amounts of intact protein can pass transcellularly and trigger mast cell activation via cross-linking of bound *IgE*. In addition, elevated *IL-4* present in individuals with an atopic background contributes towards up-regulation of the low-affinity *IgE* receptor (*CD23, FcεRII*) on the basolateral and apical poles of intestinal epithelial cells. This triggers the secretion of luminal *IgE* produced by *IL-4*-activated plasma cells; upon binding of dietary antigens, transepithelial transport back to lamina propria is initiated, leading to the passage of intact antigen capable of binding and activating mast cells (*phase I*). Upon degranulation of mast cells, mediators such as cytokines, histamine, leukotrienes (*LT*), prostaglandins (*PG*), interferon gamma (*IFN-γ*), tumour necrosis factor alpha (*TNF-α*), and monocyte chemoattractant protein (*MCP*) are released and influence ion secretion and modify paracellular permeability (*phase 2*). Alteration of the epithelial permeability occurs upon disorganization of the actinomyosin rings through activation of myosin light-chain kinase (*MLCK*) and changes in the architecture of TJs resulting from clipping of occludin by *MCP*. This leads to the entry of greater amounts of undigested allergen through the paracellular pathway, which further strengthen the intensity of the allergic reaction. (Figure is reprinted from Ref [2] with permission from John Wiley and Sons)

either the type 1 or 2 IL-4R, or of IL-13 to the type 2 IL-4R, initiates Janus kinase (Jak)-dependent tyrosine phosphorylation of the IL-4R alpha chain and the transcription factor, signal transducers and activators of transcription 6 (STAT6) [60]. It has been shown that Th2-type cytokine IL-13 influences paracellular permeability via activation of STAT6 and the increase of ions selectivity, leading to the diminished transcellular electrical resistance. Alternatively, IL-13 may activate phosphoinositol 3 kinase (PI3K) and induce apoptosis. TNF-α is also known to influence both paracellular and transcellular transport in vitro and in vivo [61, 62], as well as IL-4, another Th2 cytokine secreted by activated Th2 cells [60]. IL-4 and IL-13 increase mucosal permeability, decrease glucose absorption, and decrease chloride secretion in response to 5-hydroxytryptamine in mice's jejunum segments mounted in Ussing chambers that measure mucosal permeability. These effects were dependent on STAT6 signalling. Interestingly, responses to prostaglandin E2 (PGE2) and histamine, which were dependent on mast cells and STAT6, were enhanced by IL-4, but not by IL-13 [60]. These cytokines and enzymes secreted by activated cells combine their action and lead to a dramatical increase in epithelial permeability to food antigens. An increased amount of food allergen, which elicits an allergic response, leads to even more sever exacerbation of local, and even system allergic responses.

Studies using intestinal tissues from sensitized animals mounted in permeability-measuring chambers demonstrated that intestinal permeability changes in response to sensitizing agent. Within 2 min following the challenge, independently of presence or activation of mast cells, allergen is taken up via transcytosis by enterocytes and transported to lamina propria. The process also occurs in healthy animals, but it is enhanced in sensitized animals. Allergen transported to lamina propria encounters charged mast cells, cross-links specific IgEs, and induces release of mediators, including chymase, mast cell proteases, and histamine. Release of potent mediators disturbs epithelial junctions and allows massive entrance of allergens via the paracellular pathway. This second phase of intestinal response to an allergen occurs within 30 min following the challenge. The transcellular pathway of entrance allows intact allergens to enter into lamina propria (Fig. 2.8).

It should be noted that prior to a damage of TJs by mast cell activation, an intact allergen should have to be transported transcellularly in order to reach mast cells and provoke their activation and degranulation. It has clearly been shown that transcellular passage of intact allergens is enhanced in sensitized animals. The enhancement of the allergen transport is due to the binding to IgE expressed on the surface of epithelial cells and its transport occurs in a complex with CD23 [23]. Studies of epithelial transport in CD23 knockout mice demonstrated the significance of the enhanced transport via IgE/IgE receptor complexes [63]. Crucial role of IL-4 in enhanced expression of CD23 on enterocytes has also been demonstrated in animal studies [63]. It has also been shown that mast cells play an important role in modulating the intestinal CD23 expression and the transport of antigen/IgE/CD23 complex across human intestinal epithelial barrier [64]. Consistent with these findings, it has been demonstrated that CD23 is overexpressed in IECs of food-allergic individuals and in inflammatory bowel diseases [22]. The role of IgE in allergen transport via the complex with its receptor has also been indicated from clinical studies of intestinal secretion of allergic patients, showing elevated levels of IgE [65].

Thereby, numerous studies confirmed that food allergens were protected from lysosomal degradation, and were transported in large quantities across the epithelium by binding to the cell surface IgE/CD23 that prevented the antigenic protein from lysosomal degradation in enterocytes. However, it is not yet clear how an allergen is released from IgE/IgE receptor complex from the basolateral side of enterocytes.

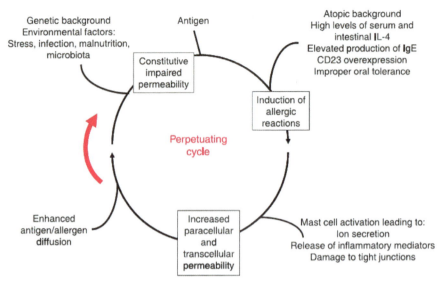

Fig. 2.8 Intestinal permeability and allergy: cause and/or consequence. Poorly defined factors associated with the genetic background or the environment may serve as priming causes to favour the entry of antigen and elicit the production of antigen-specific *IgE* in predisposed individuals. This constitutive passage will in turn promote the onset and amplification of allergy-related reactions culminating in the activation of mast cells and release of mast cell mediators with pro-inflammatory properties capable of negatively impacting the epithelial barrier function. Further damages will trigger the uncontrolled passage of more antigen, now considered as an allergen, leading to the perpetuation of the cycle allergen entry–*IgE* binding–mast cell degranulation–increased epithelial permeability. (Figure is reprinted from Ref [2] with a permission from John Wiley and Sons)

Other transport pathways may also deliver allergenic peptides and intact allergens to the basolateral side of enterocytes.

The role of IL-4 is crucial, as IL-4 not only increased production of IgE from B cells but also upregulated the expression of CD23 on IECs. The presence of IgE/CD23 opens a gate for intact dietary allergens to transcytose across the epithelial cells, and thus foments the mast cell-dependent anaphylactic responses. The understanding of the molecular mechanism responsible for epithelial barrier defects may be helpful in designing novel therapies to treat food allergy [66].

A question remains if increased epithelial permeability could contribute to food allergy development. From the available evidence, it is difficult to draw a definite conclusion. Increased permeability during the onset of a food-allergic reaction in humans does not imply that permeability was increased before allergy develops. However, both mouse and clinical studies demonstrated a link between Th2 intestinal inflammation and induction of the effector phase of food allergy.

Environmental factors that affect intestinal permeability, such as infection or stress, have consequences on susceptibility to allergic diseases [67]. Psychological stress is known to decrease mucosal barrier function [68]. Acute stress in rodents increases epithelial ionic conductance and permeation of small inert molecules (i.e., mannitol) along the paracellular pathway [69]; this stress-induced epithelial barrier defect also extends to the transcytosis of macromolecules with antigenic potential, underlining the fact that both paracellular and transcellular permeability are enhanced by stress [70, 71]. Part of the effect of stress on intestinal permeability probably derives from the release of neuroendocrine factors, such as corticotrophin-releasing hormone in the intestinal mucosa and the stimulation of cholinergic nerves [72].

IL-9-deficient mice fail to develop experimental oral antigen-induced intestinal anaphylaxis, and intestinal IL-9 overexpression induces an intestinal anaphylaxis phenotype (intestinal mastocytosis, intestinal permeability, and intravascular leakage). In addition, intestinal IL-9 overexpression predisposes to oral antigen sensitization, which requires mast cells and increased intestinal permeability [73].

There is a differential involvement of the IL-9/IL-9R pathway in systemic and oral antigen-induced anaphylaxis. Parenteral antigen-induced murine systemic anaphylaxis is mediated by both IgG- and IgE-dependent pathways, and both can occur independently of IL-9/IL-9R signalling. In contrast, oral antigen-induced intestinal and systemic anaphylaxis is strictly IgE mediated and requires IL-9/IL-9R signalling [74].

Peanut allergy in particular is among the leading causes of anaphylactic fatalities worldwide. A recent study demonstrated that peanut extract exposure in vitro induced a broad panel of responses associated with Th2/Th9-like, Th1-like, and Th17-like immunity. Peanut-dependent type 2 cytokine responses were frequently found in both peanut-allergic individuals and those who exhibit clinical tolerance to peanut ingestion. Among Th2/Th9-associated cytokines, IL-9 responses discriminated between allergic and clinically tolerant populations better than did commonly used IL-4, IL-5 or IL-13 responses. Xie et al. demonstrated that these differences were antigen dependent and allergen specific [75].

2.5 Intestinal Permeability of Modified Food Allergens

It has been demonstrated in in vitro model systems of transcytosis through a Caco-2 cell monolayer that modifications of allergens' three-dimensional structure, as well as presence of other dietary components, may influence transport and processing of allergens by enterocytes [2]. Likewise, modifications of allergens' transport and degradation during passage will also influence their capacity to activate mast cells.

Three-dimensional structure of allergens is important for its recognition by IgE antibodies against conformational epitopes of an allergen. Conformational epitopes consist of patches of protein structure, sometimes remote in sequence, which form well-defined spatial arrangement of side residues in a protein. By contrast, IgE antibodies directed against so-called linear epitopes recognized linear stretches of polypeptides. Modification of allergen's three-dimensional structure during passage and further degradation during processing will certainly destroy conformational epitopes, while linear epitopes will be less sensitive to degradation by enterocytes. These important differences between conformational and linear epitopes will

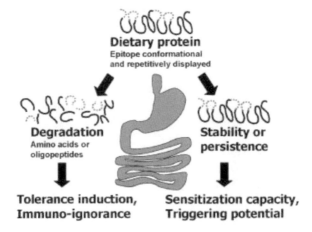

Fig. 2.9 Structural characteristics of food allergens in the sensitization and effector phase of food allergy. Degradation of dietary proteins during the gastrointestinal transit results in a hydrolysis of the proteins to free amino acids or oligopeptides, which either induce tolerance or are immunologically ignored. However, when proteins resist gastrointestinal proteases or persist during the transit due to impairment of digestion, the epitope structure remains conserved. Thus, proteins reveal sensitization potential or trigger an allergic response by intact conformation [39]

also be reflected in allergic persons' reactivity [76]. Turning this around, digested peptides do not have the capacity to sensitize and may even harbour tolerogenic properties, whereas structural motifs are strongly associated with sensitization and IgE induction.

Most of the allergic individuals who respond to conformational epitopes of food allergens will tolerate cooked, boiled, or otherwise processed food, as thermal denaturation of proteins destroys their three-dimensional structure and consequently overall conformation of allergens. By contrast, sensitivity to linear epitopes will remain even after drastic treatments of food allergens. It should also be noted that certain major food allergens show remarkable resistance to thermal denaturation or other forms of food processing due to their stable structure. Similarly, compact fold and stable structure of food allergens are often related to stability to both denaturation and proteolytic enzyme action (Fig. 2.9).

There is also a difference in clinical sensitivity among individuals recognizing conformational versus linear epitopes. Loss of reactivity to cooked food will provide partial clinical tolerance to food-allergic individuals sensitized to conformational epitopes. By contrast, reactivity to linear epitopes is often persistent during life and may result in very severe clinical manifestations of food allergy.

Numerous studies showed that antigenicity of food proteins may be influenced by modifying the conformation and structure of the protein during food processing [77, 78]. Cow's milk allergen beta-lactoglobulin (BLG) has both conformational and linear epitopes, the latter being more important in persistent forms of food allergies [79]. However, conformational change in a food protein can also expose hidden epitopes, resulting in an increase in allergenicity and antigenicity of the modified protein. Maleki et al. have shown that Maillard reactions that take place during roasting increase IgE binding [80]. It has been shown by Kleber et al. that treatment with high hydrostatic pressure enhanced the antigenic response to BLG due to unfolding and aggregation of the protein [81]. Furthermore, propensity of certain allergens to aggregate, or form, dimers has been postulated to contribute to their allergenicity. It has been shown that dimerization of food allergens increases their allergenicity, due to cross-linking of IgE-receptors by the same epitope [82]. Allergen-mediated cross-linking of IgE antibodies bound to the Fc epsilon RI receptors on the mast cell surface is the key feature of the type I allergy. If an allergen is a homodimer, its allergenicity is enhanced because it would only need one type of antibody, instead of two, for cross-linking. Numerous studies have indicated that allergens, among them food allergens, preferentially form di-, tri-, or multimers, thus leading to a repetitive display of epitopes. As B lymphocytes are pattern recognizers, this feature is essential for a memory response, but may also be critical for the very first allergen contact and initiation of the IgE response [39].

In contrast, oligomerization and polymerization of allergens by food processing may reduce the ability to efficiently cross-link IgE bound on effector cells and release of mediators of allergic inflammation. This effect is caused by a conformational change, steric hindrance, and reorientation of IgE epitopes in the highly polymerized molecule which are thus unable to cross-link FcεRI receptors on the surface of mast cells and basophils [83]. We have previously demonstrated that cross-linking of β-casein by laccase and tyrosinase causes a reduction of basophil activation in allergic patients and changes in IgE binding [84]. Similar results were obtained when BLG was treated with laccase in the presence of phenolic compound-rich sour cherry extract [85].

It has been increasingly recognized that the route by which antigen crosses the epithelium directs the outcome of the subsequent immune response. It has been proposed that during homeostasis, antigen introduced through M cells induces IgA responses, antigen delivered by goblet cell-associated antigen passages contributes to peripheral tolerance, and antigen delivered by paracellular leak initiates immune responses in the MLN [28]. Interesting observations came from the studies on processed food antigens and their route of uptake in relation to the outcome of the immune response.

Roth-Walter et al. have demonstrated that aggregation of globular cow's milk allergens BLG and alpha-lactalbumin (ALA) by pasteurization changes the uptake route in the gastrointestinal tract of animals [86]. Soluble proteins are taken up by epithelial cells, while caseins, which form micelles, are taken up by M cells of PPs. Pasteurization of milk promotes aggregation of ALA and BLG and redirects them to M cells of PPs. Pasteurization of casein does not further influence its route of uptake, or change the quality of immune response in animals. However, for BLG and ALA, pasteurization and redirection to PPs result in modification of the immune response, characterized by increased IgE and associated cytokines production, characteristic of a Th2-type response. However, aggregated proteins were less able to pass enterocytes, reach mast cells, and induce their activation. Due to reduced passage and modified structure of allergens, anaphylactic response to aggregated forms of BLG and ALA in sensitized animals was practically diminished, comparing it to the response of a sensitized animal to native, globular, allergens.

References

1. Menard S, Cerf-Bensussan N, Heyman M: **Multiple facets of intestinal permeability and epithelial handling of dietary antigens.** *Mucosal Immunol* 2010, **3**(3):247–259.
2. Perrier C, Corthesy B: **Gut permeability and food allergies.** *Clin Exp Allergy* 2011, **41**(1):20–28.
3. Suzuki T: **Regulation of intestinal epithelial permeability by tight junctions.** *Cell Mol Life Sci* 2013, **70**(4):631–659.
4. Watson CJ, Hoare CJ, Garrod DR, Carlson GL, Warhurst G: **Interferon-gamma selectively increases epithelial permeability to large molecules by activating different populations of paracellular pores.** *J Cell Sci* 2005, **118**(Pt 22):5221–5230.
5. Fihn BM, Sjoqvist A, Jodal M: **Permeability of the rat small intestinal epithelium along the villus-crypt axis: effects of glucose transport.** *Gastroenterology* 2000, **119**(4):1029–1036.
6. Colegio OR, Van Itallie CM, McCrea HJ, Rahner C, Anderson JM: **Claudins create charge-selective channels in the paracellular pathway between epithelial cells.** *Am J Physiol Cell Physiol* 2002, **283**(1):C142–147.
7. Laukoetter MG, Nava P, Lee WY, Severson EA, Capaldo CT, Babbin BA, Williams IR, Koval M, Peatman E, Campbell JA et al: **JAM-A regulates permeability and inflammation in the intestine in vivo.** *J Exp Med* 2007, **204**(13):3067–3076.
8. Vetrano S, Rescigno M, Cera MR, Correale C, Rumio C, Doni A, Fantini M, Sturm A, Borroni E, Repici A et al: **Unique role of junctional adhesion molecule-a in maintaining mucosal homeostasis in inflammatory bowel disease.** *Gastroenterology* 2008, **135**(1):173–184.
9. Vetrano S, Danese S: **The role of JAM-A in inflammatory bowel disease: unrevealing the ties that bind.** *Ann N Y Acad Sci* 2009, **1165**:308–313.
10. Assimakopoulos SF, Papageorgiou I, Charonis A: **Enterocytes' tight junctions: From molecules to diseases.** *World J Gastrointest Pathophysiol* 2011, **2**(6):123–137.
11. **Krug SM, Amasheh M, Dittmann I, Christoffel I, Fromm M, Amasheh S: Sodium caprate as an enhancer of macromolecule permeation across tricellular tight junctions of intestinal cells. Biomaterials 2013, 34(1):275–282.**
12. Kunkel D, Kirchhoff D, Nishikawa S, Radbruch A, Scheffold A: **Visualization of peptide presentation following oral application of antigen in normal and Peyer's patches-deficient mice.** *Eur J Immunol* 2003, **33**(5):1292–1301.
13. Spahn TW, Weiner HL, Rennert PD, Lugering N, Fontana A, Domschke W, Kucharzik T: **Mesenteric lymph nodes are critical for the induction of high-dose oral tolerance in the absence of Peyer's patches.** *Eur J Immunol* 2002, **32**(4):1109–1113.
14. Van Niel G, Mallegol J, Bevilacqua C, Candalh C, Brugiere S, Tomaskovic-Crook E, Heath JK, Cerf-Bensussan N, Heyman M: **Intestinal epithelial exosomes carry MHC class II/peptides able to inform the immune system in mice.** *Gut* 2003, **52**(12):1690–1697.
15. Mallegol J, Van Niel G, Lebreton C, Lepelletier Y, Candalh C, Dugave C, Heath JK, Raposo G, Cerf-Bensussan N, Heyman M: **T84-intestinal epithelial exosomes bear MHC class II/peptide complexes potentiating antigen presentation by dendritic cells.** *Gastroenterology* 2007, **132**(5):1866–1876.
16. Almqvist N, Lonnqvist A, Hultkrantz S, Rask C, Telemo E: **Serum-derived exosomes from antigen-fed mice prevent allergic sensitization in a model of allergic asthma.** *Immunology* 2008, **125**(1):21–27.
17. Prado N, Marazuela EG, Segura E, Fernandez-Garcia H, Villalba M, Thery C, Rodriguez R, Batanero E: **Exosomes from bronchoalveolar fluid of tolerized mice prevent allergic reaction.** *J Immunol* 2008, **181**(2):1519–1525.
18. Fernandez MI, Pedron T, Tournebize R, Olivo-Marin JC, Sansonetti PJ, Phalipon A: **Anti-inflammatory role for intracellular dimeric immunoglobulin a by neutralization of lipopolysaccharide in epithelial cells.** *Immunity* 2003, **18**(6):739–749.
19. Favre L, Spertini F, Corthesy B: **Secretory IgA possesses intrinsic modulatory properties stimulating mucosal and systemic immune responses.** *J Immunol* 2005, **175**(5):2793–2800.

20. Matysiak-Budnik T, Moura IC, Arcos-Fajardo M, Lebreton C, Menard S, Candalh C, Ben-Khalifa K, Dugave C, Tamouza H, van Niel G et al: **Secretory IgA mediates retrotranscytosis of intact gliadin peptides via the transferrin receptor in celiac disease.** *J Exp Med* 2008, **205**(1):143–154.

21. Heyman M, Menard S: **Pathways of gliadin transport in celiac disease.** *Ann N Y Acad Sci* 2009, **1165**:274–278.

22. Kaiserlian D, Lachaux A, Grosjean I, Graber P, Bonnefoy JY: **Intestinal epithelial cells express the CD23/Fc epsilon RII molecule: enhanced expression in enteropathies.** *Immunology* 1993, **80**(1):90–95.

23. Yang PC, Berin MC, Yu LC, Conrad DH, Perdue MH: **Enhanced intestinal transepithelial antigen transport in allergic rats is mediated by IgE and CD23(FcepsilonRII).** *J Clin Invest* 2000, **106**(7):879–886.

24. Bevilacqua C, Montagnac G, Benmerah A, Candalh C, Brousse N, Cerf-Bensussan N, Perdue MH, Heyman M: **Food allergens are protected from degradation during CD23-mediated transepithelial transport.** *Int Arch Allergy Immunol* 2004, **135**(2):108–116.

25. Li H, Nowak-Wegrzyn A, Charlop-Powers Z, Shreffler W, Chehade M, Thomas S, Roda G, Dahan S, Sperber K, Berin MC: **Transcytosis of IgE-antigen complexes by CD23a in human intestinal epithelial cells and its role in food allergy.** *Gastroenterology* 2006, **131**(1):47–58.

26. Berin MC: **Mucosal antibodies in the regulation of tolerance and allergy to foods.** *Semin Immunopathol* 2012, **34**(5):633–642.

27. Jones EA, Waldmann TA: **The mechanism of intestinal uptake and transcellular transport of IgG in the neonatal rat.** *Gut* 1971, **12**(10):855–856.

28. Knoop KA, Miller MJ, Newberry RD: **Transepithelial antigen delivery in the small intestine: different paths, different outcomes.** *Curr Opin Gastroenterol* 2013, **29**(2):112–118.

29. Diehl GE, Longman RS, Zhang JX, Breart B, Galan C, Cuesta A, Schwab SR, Littman DR: **Microbiota restricts trafficking of bacteria to mesenteric lymph nodes by CX(3)CR1(hi) cells.** *Nature* 2013, **494**(7435):116–120.

30. McDole JR, Wheeler LW, McDonald KG, Wang B, Konjufca V, Knoop KA, Newberry RD, Miller MJ: **Goblet cells deliver luminal antigen to CD103+ dendritic cells in the small intestine.** *Nature* 2012, **483**(7389):345–349.

31. Kim YS, Ho SB: **Intestinal goblet cells and mucins in health and disease: recent insights and progress.** *Curr Gastroenterol Rep* 2010, **12**(5):319–330.

32. Macierzanka A, Bottger F, Lansonneur L, Groizard R, Jean AS, Rigby NM, Cross K, Wellner N, Mackie AR: **The effect of gel structure on the kinetics of simulated gastrointestinal digestion of bovine beta-lactoglobulin.** *Food Chem* 2012, **134**(4):2156–2163.

33. Troncoso E, Aguilera JM, McClements DJ: **Influence of particle size on the in vitro digestibility of protein-coated lipidnanoparticles.** *J Colloid Interface Sci* 2012, **382**(1):110–116.

34. Furlund CB, Ulleberg EK, Devold TG, Flengsrud R, Jacobsen M, Sekse C, Holm H, Vegarud GE: **Identification of lactoferrin peptides generated by digestion with human gastrointestinal enzymes.** *J Dairy Sci* 2013, **96**(1):75–88.

35. Macierzanka A, Bottger F, Rigby NM, Lille M, Poutanen K, Mills EN, Mackie AR: **Enzymatically structured emulsions in simulated gastrointestinal environment: impact on interfacial proteolysis and diffusion in intestinal mucus.** *Langmuir* 2012, **28**(50):17349–17362.

36. Girgih AT, Udenigwe CC, Aluko RE: **Reverse-phase HPLC Separation of Hemp Seed (Cannabis sativa L.) Protein Hydrolysate Produced Peptide Fractions with Enhanced Antioxidant Capacity.** *Plant Foods Hum Nutr* 2013.

37. Prandi B, Farioli L, Tedeschi T, Pastorello EA, Sforza S: **Simulated gastrointestinal digestion of Pru ar 3 apricot allergen: assessment of allergen resistance and characterization of the peptides by ultra-performance liquid chromatography/electrospray ionisation mass spectrometry.** *Rapid Commun Mass Spectrom* 2012, **26**(24):2905–2912.

38. Astwood JD, Leach JN, Fuchs RL: **Stability of food allergens to digestion in vitro.** *Nat Biotechnol* 1996, **14**(10):1269–1273.

39. Untersmayr E, Jensen-Jarolim E: **Mechanisms of type I food allergy.** *Pharmacol Ther* 2006, **112**(3):787–798.
40. Mowat AM: **Dendritic cells and immune responses to orally administered antigens.** *Vaccine* 2005, **23**(15):1797–1799.
41. Rimoldi M, Rescigno M: **Uptake and presentation of orally administered antigens.** *Vaccine* 2005, **23**(15):1793–1796.
42. Bilsborough J, Viney JL: **Gastrointestinal dendritic cells play a role in immunity, tolerance, and disease.** *Gastroenterology* 2004, **127**(1):300–309.
43. Allez M, Brimnes J, Dotan I, Mayer L: **Expansion of CD8+ T cells with regulatory function after interaction with intestinal epithelial cells.** *Gastroenterology* 2002, **123**(5):1516–1526.
44. Farhadi A, Banan A, Fields J, Keshavarzian A: **Intestinal barrier: an interface between health and disease.** *J Gastroenterol Hepatol* 2003, **18**(5):479–497.
45. Mishra A, Makharia GK: **Techniques of functional and motility test: how to perform and interpret intestinal permeability.** *J Neurogastroenterol Motil* 2012, **18**(4):443–447.
46. Johnston SD, Smye M, Watson RP: **Intestinal permeability tests in coeliac disease.** *Clin Lab* 2001, **47**(3–4):143–150.
47. Ukabam SO, Clamp JR, Cooper BT: **Abnormal small intestinal permeability to sugars in patients with Crohn's disease of the terminal ileum and colon.** *Digestion* 1983, **27**(2):70–74.
48. Juby LD, Rothwell J, Axon AT: **Cellobiose/mannitol sugar test–a sensitive tubeless test for coeliac disease: results on 1010 unselected patients.** *Gut* 1989, **30**(4):476–480.
49. Dupont C, Barau E, Molkhou P, Raynaud F, Barbet JP, Dehennin L: **Food-induced alterations of intestinal permeability in children with cow's milk-sensitive enteropathy and atopic dermatitis.** *J Pediatr Gastroenterol Nutr* 1989, **8**(4):459–465.
50. Andre C, Andre F, Colin L, Cavagna S: **Measurement of intestinal permeability to mannitol and lactulose as a means of diagnosing food allergy and evaluating therapeutic effectiveness of disodium cromoglycate.** *Ann Allergy* 1987, **59**(5 Pt 2):127–130.
51. Ventura MT, Polimeno L, Amoruso AC, Gatti F, Annoscia E, Marinaro M, Di Leo E, Matino MG, Buquicchio R, Bonini S *et al*: **Intestinal permeability in patients with adverse reactions to food.** *Dig Liver Dis* 2006, **38**(10):732–736.
52. Laudat A, Arnaud P, Napoly A, Brion F: **The intestinal permeability test applied to the diagnosis of food allergy in paediatrics.** *West Indian Med J* 1994, **43**(3):87–88.
53. Cencic A, Langerholc T: **Functional cell models of the gut and their applications in food-microbiology–a review.** *Int J Food Microbiol* 2010, **141** Suppl 1:S4–14.
54. Jackson PG, Lessof MH, Baker RW, Ferrett J, MacDonald DM: **Intestinal permeability in patients with eczema and food allergy.** *Lancet* 1981, **1**(8233):1285–1286.
55. Scudamore CL, Thornton EM, McMillan L, Newlands GF, Miller HR: **Release of the mucosal mast cell granule chymase, rat mast cell protease-II, during anaphylaxis is associated with the rapid development of paracellular permeability to macromolecules in rat jejunum.** *J Exp Med* 1995, **182**(6):1871–1881.
56. Scudamore CL, Jepson MA, Hirst BH, Miller HR: **The rat mucosal mast cell chymase, RMCP-II, alters epithelial cell monolayer permeability in association with altered distribution of the tight junction proteins ZO-1 and occludin.** *Eur J Cell Biol* 1998, **75**(4):321–330.
57. McDermott JR, Bartram RE, Knight PA, Miller HR, Garrod DR, Grencis RK: **Mast cells disrupt epithelial barrier function during enteric nematode infection.** *Proc Natl Acad Sci U S A* 2003, **100**(13):7761–7766.
58. Ma TY, Boivin MA, Ye D, Pedram A, Said HM: **Mechanism of TNF-{alpha} modulation of Caco-2 intestinal epithelial tight junction barrier: role of myosin light-chain kinase protein expression.** *Am J Physiol Gastrointest Liver Physiol* 2005, **288**(3):G422–430.
59. Weber CR, Raleigh DR, Su L, Shen L, Sullivan EA, Wang Y, Turner JR: **Epithelial myosin light chain kinase activation induces mucosal interleukin-13 expression to alter tight junction ion selectivity.** *J Biol Chem* 2010, **285**(16):12037–12046.
60. Madden KB, Whitman L, Sullivan C, Gause WC, Urban JF, Jr., Katona IM, Finkelman FD, Shea-Donohue T: **Role of STAT6 and mast cells in IL-4- and IL-13-induced alterations in murine intestinal epithelial cell function.** *J Immunol* 2002, **169**(8):4417–4422.

61. Wang F, Graham WV, Wang Y, Witkowski ED, Schwarz BT, Turner JR: **Interferon-gamma and tumor necrosis factor-alpha synergize to induce intestinal epithelial barrier dysfunction by up-regulating myosin light chain kinase expression**. *Am J Pathol* 2005, **166**(2):409–419.
62. Clayburgh DR, Musch MW, Leitges M, Fu YX, Turner JR: **Coordinated epithelial NHE3 inhibition and barrier dysfunction are required for TNF-mediated diarrhea in vivo**. *J Clin Invest* 2006, **116**(10):2682–2694.
63. Yu LC, Yang PC, Berin MC, Di Leo V, Conrad DH, McKay DM, Satoskar AR, Perdue MH: **Enhanced transepithelial antigen transport in intestine of allergic mice is mediated by IgE/CD23 and regulated by interleukin-4**. *Gastroenterology* 2001, **121**(2):370–381.
64. Tu YH, Oluwole C, Struiksma S, Perdue MH, Yang PC: **Mast cells modulate transport of CD23/IgE/antigen complex across human intestinal epithelial barrier**. *N Am J Med Sci* 2009, **1**(1):16–24.
65. Belut D, Moneret-Vautrin DA, Nicolas JP, Grilliat JP: **IgE levels in intestinal juice**. *Dig Dis Sci* 1980, **25**(5):323–332.
66. Yu LC: **The epithelial gatekeeper against food allergy**. *Pediatr Neonatol* 2009, **50**(6):247–254.
67. Heyman M: **Gut barrier dysfunction in food allergy**. *Eur J Gastroenterol Hepatol* 2005, **17**(12):1279–1285.
68. Meddings JB, Swain MG: **Environmental stress-induced gastrointestinal permeability is mediated by endogenous glucocorticoids in the rat**. *Gastroenterology* 2000, **119**(4):1019–1028.
69. Saunders PR, Kosecka U, McKay DM, Perdue MH: **Acute stressors stimulate ion secretion and increase epithelial permeability in rat intestine**. *Am J Physiol* 1994, **267**(5 Pt 1):G794–799.
70. Gareau MG, Silva MA, Perdue MH: **Pathophysiological mechanisms of stress-induced intestinal damage**. *Curr Mol Med* 2008, **8**(4):274–281.
71. Kiliaan AJ, Saunders PR, Bijlsma PB, Berin MC, Taminiau JA, Groot JA, Perdue MH: **Stress stimulates transepithelial macromolecular uptake in rat jejunum**. *Am J Physiol* 1998, **275**(5 Pt 1):G1037–1044.
72. Santos J, Saunders PR, Hanssen NP, Yang PC, Yates D, Groot JA, Perdue MH: **Corticotropin-releasing hormone mimics stress-induced colonic epithelial pathophysiology in the rat**. *Am J Physiol* 1999, **277**(2 Pt 1):G391–399.
73. Forbes EE, Groschwitz K, Abonia JP, Brandt EB, Cohen E, Blanchard C, Ahrens R, Seidu L, McKenzie A, Strait R et al: **IL-9- and mast cell-mediated intestinal permeability predisposes to oral antigen hypersensitivity**. *J Exp Med* 2008, **205**(4):897–913.
74. Osterfeld H, Ahrens R, Strait R, Finkelman FD, Renauld JC, Hogan SP: **Differential roles for the IL-9/IL-9 receptor alpha-chain pathway in systemic and oral antigen-induced anaphylaxis**. *J Allergy Clin Immunol* 2010, **125**(2):469–476 e462.
75. Xie J, Lotoski LC, Chooniedass R, Su RC, Simons FE, Liem J, Becker AB, Uzonna J, Hay-Glass KT: **Elevated antigen-driven IL-9 responses are prominent in peanut allergic humans**. *PLoS One* 2012, **7**(10):e45377.
76. Vila L, Beyer K, Jarvinen KM, Chatchatee P, Bardina L, Sampson HA: **Role of conformational and linear epitopes in the achievement of tolerance in cow's milk allergy**. *Clinical and Experimental Allergy* 2001, **31**(10):1599–1606.
77. Tedford LA, Kelly SM, Price NC, Schaschke CJ: **Combined effects of thermal and pressure processing on food protein structure**. *Food Bioprod Process* 1998, **76**(C2):80–86.
78. Sathe SK, Teuber SS, Roux KH: **Effects of food processing on the stability of food allergens**. *Biotechnology Advances* 2005, **23**(6):423–429.
79. Ball G, Shelton MJ, Walsh BJ, Hill DJ, Hosking CS, Howden ME: **A major continuous allergenic epitope of bovine beta-lactoglobulin recognized by human IgE binding**. *Clin Exp Allergy* 1994, **24**(8):758–764.
80. Maleki SJ, Chung SY, Champagne ET, Raufman JP: **The effects of roasting on the allergenic properties of peanut proteins**. *J Allergy Clin Immunol* 2000, **106**(4):763–768.
81. Kleber N, Krause I, Illgner S, Hinrichs J: **The antigenic response of beta-lactoglobulin is modulated by thermally induced aggregation**. *Eur Food Res Technol* 2004, **219**(2):105–110.

82. Rouvinen J, Janis J, Laukkanen ML, Jylha S, Niemi M, Paivinen T, Makinen-Kiljunen S, Haahtela T, Soderlund H, Takkinen K: **Transient Dimers of Allergens**. *Plos One* 2010, **5**(2):e9037.
83. Niederberger V: **Allergen-specific immunotherapy**. *Immunol Lett* 2009, **122**(2):131–133.
84. Stanic D, Monogioudi E, Dilek E, Radosavljevic J, Atanaskovic-Markovic M, Vuckovic O, Lannto R, Mattinen M, Buchert J, Cirkovic Velickovic T: **Digestibility and allergenicity assessment of enzymatically crosslinked beta-casein**. *Mol Nutr Food Res* 2010, **54**(9):1273–1284.
85. Tantoush Z, Stanic D, Stojadinovic M, Ognjenovic J, Mihajlovic L, Atanaskovic-Markovic M, Cirkovic Velickovic T: **Digestibility and allergenicity of beta-lactoglobulin following laccase-mediated cross-linking in the presence of sour cherry phenolics**. *Food Chem* 2011, **125**(1):84–91.
86. Roth-Walter F, Berin MC, Arnaboldi P, Escalante CR, Dahan S, Rauch J, Jensen-Jarolim E, Mayer L: **Pasteurization of milk proteins promotes allergic sensitization by enhancing uptake through Peyer's patches**. *Allergy* 2008, **63**(7):882–890.

Chapter 3
Biochemistry and Molecular Biology of Food Allergens

Contents

Abbreviations

3D	Three-dimensional structure
ABPs	Auxin-binding proteins
ALA	α-lactalbumin
BAT	Basophil activation test
BLG	β-lactoglobulin
CD	Clusted of differentiation
CRD	Component-resolved diagnosis
CSA	Chicken serum albumin
DBPCFC	Double-blind placebo-controlled food challenge
EAST	Enzyme Allergo-Sorbent Testing
ELISA	Ezyme-Linked Immunosorbent Assay
GLPs	Germins and germin-like proteins
ISAC	Immuno Solid-phase Allergen Chip
nsLTPs	Nonspecific lipid transfer proteins
OAS	Oral allergy syndrome
PR	Pathogenesis related
RAST	Radioallergosorbent test

T. Ćirković Veličković, M. Gavrović-Jankulović, *Food Allergens*,
Food Microbiology and Food Safety, DOI 10.1007/978-1-4939-0841-7_3,
© Springer Science+Business Media New York 2014

SPT   Skin prick tests
VTG   Vitellogenin
WHO/IUIS   World Health Organization/International Union of Immunological
            Societies
YGP   Yolk glycoprotein

Summary A food allergen has the ability to first elicit an immunoglobulin E (IgE) response, and then, on subsequent exposures, to elicit a clinical response to the same or similar protein. However, despite increasing knowledge of the structure and amino acid sequences of the identified allergens, only a few biochemical characteristics can be associated with food allergens, such as the biochemical characteristics that allow a food protein to survive the extremes of food processing, escape the digestive enzymes of the human gastrointestinal tract, and interact with the immune system. Food allergen characteristics include abundance of the protein in the food, presence of multiple linear IgE binding epitopes, and resistance of the protein to digestion and processing.

Most plant and animal food allergens belong to only several protein families. It seems likely that the bulk of protein families that include allergenic members have already been discovered. However, belonging to one of a limited number of protein families is not sufficient to determine allergenic activity of a given protein. It seems that the factors important in sensitization of an atopic individual with any given allergen are: (1) the genetic predisposition of the exposed person, (2) the structure of the allergen, and (3) the biochemical and physicochemical properties of the allergen.

Various factors may facilitate the presentation of food allergens to the immune system (primarily through the gut) and these include stability enhanced by binding various types of ligands, large number of disulfide bonds, oligomerization or aggregation, glycosylation, and potential interaction with cell membranes or lipid structures. Some biochemical characteristics associated with food allergens, such as the presence of multiple, linear IgE-binding epitopes and the resistance of the protein to digestion and processing, seem to predominate among food allergens, more so than common structural features.

3.1   Characterization of Food Allergens

A large number of food allergens able to induce allergic symptoms in predisposed individuals, including severe, even life-threatening reactions, have been identified and characterized. However, proteins able to induce such immunoglobulin E (IgE)-mediated reactions can be assigned to only a limited number of protein families. Detailed knowledge on the characteristics of food allergens, and their three-dimensional (3D) structure, biological activity, and stability, will help to improve diagnosis of food allergy, avoid unnecessary exclusion diets, and assess the risk of cross-reactive allergies to other food sources [1].

Various in vivo and in vitro assays have been developed for assessment of IgE reactivity of allergenic proteins form foods.

In Vivo Tests In vivo detection of IgE reactivity includes skin prick tests (SPT) and double-blind placebo-controlled food challenge (DBPCFC). SPT is a simple test, but the wheal sizes can vary by allergen and subject [2], subjective results can differ between evaluators, and patients with atopic dermatitis may develop false-positive wheals [3]. The DBPCFC is universally accepted as the gold standard of food allergy diagnosis, because of the high probability of false-positive results of open food challenges. However, these tests are complex, expensive, and time consuming. Patients who are susceptible to anaphylaxis should not be included in this type of study [3]. The main disadvantage of the DBPCFC is that it is labor intensive, time consuming, and therefore expensive.

Animal Models Rodent animal models are the most often exploited animal models of food allergy; however, mouse models have also been developed [4–6]. Animal models are useful in elucidation of mechanisms involved in reduction of allergenicity [7–9].

Ex Vivo Tests Ex vivo tests include measurement of histamine release or upregulation of surface molecules CD63 or CD203c on basophil granulocytes, known as the basophil activation test (BAT). Practical application of BAT has limitations, such as the availability of basophil donors and time constraints.

In Vitro Tests In vitro studies for determining allergen reactivity include the measurement of serum-specific IgE using various types of quantitative (radioallergosorbent test, RAST; enzyme allergosorbent testing, EAST; enzyme-linked immunosorbent assay, ELISA; ImmunoCAP assays) and semiquantitative immunoassays (immunoblot, dot blot).

The *mediator release assay* has been developed for in vitro assessment of biological potency of allergens and allergen extracts. In this assay, the rat basophilic leukemia (RBL) cell line transfected with human Fcε receptor type1 is passively sensitized with human IgE. The effect of the interaction between the allergen and allergen-specific IgE is demonstrated via mediator release. RBL cells have the functional characteristics of mast cells with regard to IgE-induced mediator release. Allergen detection levels can be in the range of pg/mL. However, there are subjects with detectable levels of specific IgE (sIgE) that do not induce good mediator release in the mediator release assay (nonresponders) [10].

Discrepancies between the magnitude of mediator release in the in vitro assay and serum allergen-specific IgE have been reported [11]. Lack of responsiveness in the mediator release assay despite detectable allergen-specific IgE has been attributed to: (1) low affinity and avidity of IgE [12], (2) effect of the steric site of allergen recognition [13, 14], (3) availability of free (not bound in complexes) IgE [15, 16], and (4) low ratio of allergen-specific IgE to total serum IgE (dilution effect) [17].

During the past decades, a new in vitro technique, the so-called component-resolved diagnosis (CRD), has entered the field of allergy. In contrast to traditional sIgE assays, CRD does not rely upon whole extract preparations from native allergens but on quantification of sIgE antibodies to single protein components, purified from natural sources or obtained by recombinant DNA technology [18]. It emerges that CRD can improve management of the allergic patients, as it allows, to a certain extent, discrimination between clinically significant and irrelevant sIgE results, and allows the establishment of sensitization patterns with particular prognostic outcomes and therapeutic approaches. However, in spite of enormous progress made in the previous years in allergen identification and characterization, it is very likely that all IgE reactive components from various natural source materials have not been identified yet, and such IgE reactivity would be overlooked by this approach. Nevertheless, the CRD concept, coined by Rudolf Valenta [19], together with advancements made in the past decades in biochip technology, enabled the development of a novel diagnostic tool for high-resolution IgE profiling—allergen microarrays. The advanced in vitro diagnostic test for simultaneous measurement of sIgE antibodies to a broad spectrum of allergen components, the ImmunoCAP® ISAC, has been developed. Basically, it is a semiquantitative microarray-based solid-phase immunoassay in which the allergen components (high-quality recombinant and purified) are spotted in triplets and covalently immobilized to a polymer-coated slide. Minute amounts of serum or plasma are necessary (30 µL) for the test. The assay is performed in two steps: the first is binding of IgE antibodies from the patient sample to the immobilized allergen components, and the second is detection of allergen-bound IgE by a fluorescence-labeled anti-IgE. Results are reported in ISAC Standardized Units (ISU) giving indications of sIgE antibody levels. The whole test procedure gives a total assay time of less than 4 h. Clinical usefulness of microarray-based IgE detection, especially in children with suspected food allergy, has been shown [20]. Allergen microarray has been recognized as a useful tool to diagnose symptomatic cow's milk and hen egg allergy in the pediatric population, and shows performance characteristics comparable to the current diagnostic tests. The advantage of this microarray method is that it may be used to indicate allergy in small children in whom only small blood volumes are obtainable. However, the assay is in most cases not capable of replacing DBPCFC. In addition, one may expect identification of novel food allergens which should then be also implemented in the assay.

3.2 Plant Food Allergen Families

Plant food allergens were classified on the basis of their biologic function, or by the membership to protein families [21]. According to the Pfam protein database, all plant food allergens fall into 31 of 8,296 protein families [22]. The most important plant food allergens can be grouped into three big protein families: the prolamin and cupin superfamilies, and the family 10 of pathogenesis-related proteins (PRs) [23]. The prolamin, cupin, profilin, and PR-10 protein families contain approximately 65 % of all plant food allergens, while the remaining 27 allergen-containing protein families refer to mostly the plant defense system or PRs.

3.2.1 The Prolamin Family

The *prolamin* superfamily comprises the largest number of allergenic plant food proteins. Prolamins are proline- and glutamine-rich proteins with a conserved skeleton of eight cysteine residues that play several biological functions. They comprise three major groups of plant food allergens: the *seed storage 2S albumins* found in tree nuts and seeds, the defensin-related *nonspecific lipid transfer proteins* (nsLTPs) found in soft fruits and vegetables, and cereal α-*amylase/trypsin inhibitors* (Table 3.1) [21, 24]. Allergies to cereal prolamin seed storage proteins do not occur very frequently and have been studied mainly in wheat [25]. These protein families have little sequence homology to each other apart from the cysteine skeleton, but they have highly similar α-helical structure, which is highly stable to both thermal and proteolytic denaturation and might contribute to the allergenicity of these proteins [26].

2S albumins from different plant families show sequence identities of less than 40 %. However, despite low general sequence similarity, high sequence similarity between linear IgE epitopes was shown for 2S albumins from cashew and walnut [27]. In this regard, prediction of cross-reactivity based on the global sequence similarity among food allergens does not seem to be a reliable parameter, as they very often undergo partial denaturation during the digestive process [28].

3.2.2 The Cupin Family

The *cupins* are the second major superfamily of plant food allergens, which are widely distributed among all kingdoms and share a conserved six-stranded β-barrel fold [29]. However, allergenicity within the cupins is confined to the vicilin and legumin seed storage proteins. The cupin superfamily comprises the allergenic 7S and 11S globulin storage proteins from peanuts, soybean, and tree nuts (Table 3.1), and these proteins are heat stable [21, 30].

Cupin subgroups are classified by whether proteins have a single cupin domain or whether they have a duplicated or a multicupin structure. It has been estimated that there are a minimum of 18 different functional subclasses based on the various enzymatic and nonenzymatic functions of the proteins [31].

Monocupins (single-domain cupins) comprise the majority of cupin proteins and are monomeric, dimeric, or oligomeric. They mostly comprise enzymes, such as dioxygenases and phosphomannose isomerases [31]. Auxin-binding proteins (ABPs), a family of dimeric monocupins from plants, are involved in a variety of plant growth responses by interacting with the plant hormone auxin [32]. ABPs display considerable sequence conservation across a broad range of plants, including apple and strawberry [23].

Germins and germin-like proteins (GLPs), being oligomeric monocupins, represent the largest family of cupins in plants. Germins from wheat and barley are highly thermostable, hydrogen peroxide-generating oxalate oxidases, with an additional superoxide dismutase activity identified for barely germin [32].

Table 3.1 The main plant food allergens

Allergen name	Botanical source	Protein family	Function in plant	MM (kDa)	Biochemical characteristics	Reference
α-gliadin γ-gliadin LMW glutenin	Wheat (*Triticum aestivum*)	Prolamin superfamily Prolamin superfamily Prolamins	Seed storage protein	31–42 33–44	They are rich in proline and glutamine. Only certain members of this multigene family are allergens	[41–46]
Pru p 1 Mal d 3 Zea m 14 Pru ar 3 Pru av 3	Peach (*Prunus persica*) Apple (*Malus domesticus*) Maize (*Zea mays*) Apricot (*Prunus armeniaca*) Cherry (*Prunus avium*)	Nonspecific lipid transfer proteins (nsLTPs)	Function uncertain; maybe involved in transport of suberin monomers	9 11.4 9 9 9	Members of the nsLTP family are thermo-stable. They are also resistant to digestion. No glycosylation has been observed in plants	[47, 48] [49–50] [51] [52] [53]
Hor v 1 Rag 1, 2, 5, 5b, 14, 14b, 16, 17	Wheat (*Triticum aestivum*) Barley (*Hordeum vulgarum*) Maize (*Zea mays*) Rice (*Oryza sativa*)	α-amylase/trypsin inhibitors	Provide protection against degradative proteasese and amylases produced by insect pests and pathogens	15 16 16 14–16	They appear to be thermo-stable, retaining their allergenicity following processes such as brewing	[54] [55] [51] [56–58]
Jug r 1 Sin a 1 Bra j 1 Ber e 1 Ara h 2 Ara h 6 Ara h 7 Ses i 1 SFA-8	Walnut (*Juglans regia*) Yellow mustard (*Sinapis alba*) Oriental mustard (*Brassica juncea*) Brazil nut (*Bertolletia excelsa*) Peanut (*Arachis hypogea*) Sesame (*Sesamum indicum*) Sunflower (*Helianthus annus*)	2S albumins	Seed storage proteins	16 14 14.7 12 17.5 14.5 15.8 9 12	Allergen present in raw and roasted seeds. Thermostability has not been studied in detail but they are highly resistant to proteolysis. They may be glycosylated in certain plant species	[59] [60] [61] [62] [63–65] [66] [67]

Table 3.1 (continued)

Allergen name	Botanical source	Protein family	Function in plant	MM (kDa)	Biochemical characteristics	Reference
Ara h 1 Jug r 2 β-Conglycinin	Peanut (*Arachis hypogea*) Walnut (*Juglans regia*) Soya (*Glycine max*)	Cupin superfamily 7S (vicilin-like) globulins	Seed storage protein	60 47 60	Proteins comprise 50–60,000-Da subunits assembled into trimers. The trimers may be glycosylated. The trimer is resistant to pepsinolysis. Proteins are thermostable up to 75°C	[63, 68] [59] [69]
Ara h 3 Ara h 4 Glycinin	Peanut (*Arachis hypogea*) Soy (*Glycine max*)	11S (legumin-like) globulins	Seed storage protein	61 54	These proteins are thermostable up to 92–95°C. Form intermediates on trypsinolysis which retain much of the quaternary structure. They are rarely glycosylated	[70] [71, 72]
Act d 1 (actinidin) Gly m Bd 30k P34	Kiwi fruit (*Actinidia chinensis*) Soy (*Glycine max*)	Cysteine protease C 1 family	Cysteine proteases	30 34	Synthesised as a precursor protein, some are *N*-glycosylated	[73] [74]

The first discovered two-domain cupins, or bicupins, were the seed storage proteins of higher plants [33]. They can be classified into trimeric 7S globulins or *vicilins* and hexameric 11S globulins or *legumins,* to which peanut and tree nut allergens belong, such as Ara h 1 from peanut and Jug r 2 from walnut. 7S and 11S globulins share relatively low sequence identities of 35–45 % but display high structural similarity [34, 35]. The cupin core provides a stable scaffold that allows these proteins to survive and function under a great variety of extreme conditions [36]. The majority of cupin allergens belongs to either the vicilin-like or the legumin-like seed storage protein families and comprise major legume, tree nut, and seed allergens [21]. GLPs have been described as allergens in pepper, orange, and tangerine, owing much of their IgE reactivity to their glycosylation [37].

3.2.3   The Profilin and Bet v 1 Family

The *profilin* and *Bet v 1 family* includes tree pollinosis-associated food allergens with low stability that induce symptoms of oral allergy.

Profilins, cytosolic proteins of 12–15 kDa, are found in all eukaryotic cells. Profilins bind to monomeric actin and a plethora of other ligands (i.e., phosphatidyl inositol phosphate) and diverse regulatory proteins that contain proline-rich stretches. Profilin is regarded as a key player in the regulation of actin cytoskeleton dynamics during processes such as cell movement, cytokinesis, and signaling. Although the sequence similarity between profilins from vertebrate and other organisms is low, 3D structures of all profilins are strikingly similar. They fold into compact globular mixed α/β structures. Because of their great extent of sequence conservation, they constitute a family of highly cross-reactive allergens in monocot and dicot pollens, plant foods, and *Hevea latex* [23].

Many of the known plant food allergens are homologous to PRs, proteins that are induced by pathogens or certain environmental stresses. PRs have been classified into 17 families [38]. Allergens homologous to PRs include chitinases (PR-3 family), antifungal proteins such as the thaumatin-like proteins (PR-5), proteins homologous to the major birch pollen allergen Bet v 1 (PR-10) from vegetables and fruits and lipid transfer proteins (PR-14) from fruits and cereals [39].

Allergens homologous to Bet v 1 (major birch pollen allergen) constitute a group of PRs (PR-10), with a molecular weight of 17 kDa (Table 3.1), which behave as major allergens in patients with allergy to vegetables associated with birch pollen allergy. In these patients, the primary sensitization seems to be produced through the inhalation route on exposure to birch pollen with symptomatology of oral allergy syndrome (OAS). Although Bet v 1 is the primary sensitizing allergen in allergies to foods containing Bet v 1 homologues, continuous exposure to carrots, however, results in recognition of discrete epitopes on the major allergen Dau c 1. The Be v 1 homologous proteins are not characterized by extreme stability, as many other food allergens are [23, 39, 40].

3.3 Animal Food Allergens

Animal food allergens have a lower diversity than those found in plants. They can be classified into three main families—EF hand proteins (parvalbumin), tropomyosins (crustaceans and mollusks), and caseins—together with a long tail of families containing only a few reported allergens in each [75]. For all three main animal allergen families, their ability to act as allergens seems to be related to their closeness to human homologues. Interestingly, proteins with a sequence identity of 54% to human homologues are all allergenic, whereas those with a sequence identity greater than 63% to human homologues were rarely allergenic. This observation probably relates to the requirement for proteins to be recognized as nonself to mount an immune response, and it has been argued that a low degree of similarity to a host's proteome is required for immunogenicity [76]. The most important animal food allergens are present in milk, egg, and seafood. Their properties are shown in Table 3.2.

3.3.1 Parvalbumin and Fish Allergens

Parvalbumins represent the biggest group of animal-derived food allergens; they belong to the EF hand domain family (http://www.meduniwien.ac.at/allergens/allfam/) and contain more than 63 allergens reported by far [77]. Parvalbumin is resistant to heat, chemical denaturation, and proteolytic enzymes [78]. The main function of parvalbumin is in the muscle contraction/relaxation cycle, calcium buffering, and signal transduction. Parvalbumins are typically 10–12 kDa in size and acidic (pI=4.0–5.2). They are structurally characterized by the presence of three typical helix–loop–helix domains (EF hand domain), two of which are able to bind divalent cations, like Ca^{2+}[79]. Parvalbumin is the major clinical cross-reactive fish allergen, and 90% of fish-allergic patients react to this protein [80–82]. Parvalbumins can be found in α and β isoform lineages. Fish often contain both α and β parvalbumin; however, the majority of allergenic parvalbumins reported belong to the β lineage. Most fish express two or more different β parvalbumin isoforms, which are designated β1, β2, and so on [83]. These β isoforms can differ significantly in amino acid sequence as demonstrated for Atlantic salmon (*Salmo salar*) where their β1 and β2 isoforms have more than 64% identity. The differences in β parvalbumin isoforms in one species can result in a fish-allergic patient reacting to one isoform more than another, which contributes to the complexity of diagnosing fish allergy and detection of allergenic parvalbumin [84]. Dimeric and polymeric forms of parvalbumin have been reported to bind IgE and these allergens form higher molecular mass aggregates of about 24 and 48 kDa [85, 86]. The allergenicity of parvalbumin has been studied in a number of fish species and as of 2012, the allergome database (www.allergome.org) has 218 allergenic isoforms of fish parvalbumin listed.

Besides β-parvalbumin (Gad m 1) in Atlantic cod (*Gadus morhua*) enzymes such as β-enolase (47.3 kDa, Gad m 2) [87] and aldolase A (40 kDa, Gad m 3) [88] have

Table 3.2 The main animal food allergens

Allergen name	Allergen source	Protein family	MM (kDa)	Reference
Gad m 1(parvalbumin)	Atlantic cod (*Gadus morhua*)	EF hand domain Ca-binding protein	12	[83–86]
Gad m 2	Atlantic cod (*Gadus morhua*)	Beta-enolase	47.3	[87]
Gad m 3	Atlantic cod (*Gadus morhua*)	Aldolase A	40	[88]
Pen m 1 (tropomyosin)	Shrimp (*Penaeus monodon*)	Tropomyosins	34–38	[94, 95]
Pen m 2	Shrimp (*Penaeus monodon*)	Arginine kinases	40	[96]
Pen m 3 (myosin-like protein)	Shrimp (*Penaeus monodon*)	Myosins	19	[97]
Pen m 4	Shrimp (*Penaeus monodon*)	Sarcoplasmic calcium-binding protein	22	[97]
Pen m 6	Shrimp (*Penaeus monodon*)	Troponin C	17	[98]
Bos d 4 (α-lactalbumin)	Milk (*Bos domesticus*)	α-lactalbumin	14.2	[105]
Bos d 5 (β-lactoglobulin)	Milk (*Bos domesticus*)	Lipocalin	18.3	[103]
Bos d 6 (serum albumin)	Milk (*Bos domesticus*)	Serum albumin	67	[101]
Bos d 7 (immunoglobulin)	Milk (*Bos domesticus*)	Immunoglobulins	160	[121]
Bos d 8 (casein)	Milk (*Bos domesticus*)	Caseins	20–30	[122]
Gal d 1 (ovomucoid, trypsin inhibitor)	Egg (*Gallusdomesticus*)	Kazal-type serine protease inhibitors	28	[106, 107]
Gal d 2 ovalbumin	Egg (*Gallusdomesticus*)	Serine-type endopeptidase inhibitor	43	[112–114]
Gal d 3 (conalbumin, ovotransferrin)	Egg (*Gallusdomesticus*)	Transferrins	78	[111, 115]
Gal d 4 (lysozyme)	Egg (*Gallusdomesticus*)	c-type lysozyme	16.2	[111, 117]

been identified. Enzymes such as β-enolase (50 kDa, Thu a 2) and aldolase (40 kDa, Thu a 3) from yellowfin tuna (*Thunnus albacares*) and β-enolase (47.2 kDa, Sal s 2) and aldolase A (40 kDa, Sal s 3) from Atlantic salmon (*Salmo salar*) have also been registered as allergens by WHO/IUIS (www.allergen.org).

Besides allergens derived from fish themselves, contaminants such as the parasite *Anisakis simplex* can cause allergic reactions [89]. Among 12 identified allergens in this parasite are tropomyosin (Ani s 3), paramyosin (Ani s 2), cysteine protease inhibitor (Ani s 4), and serine protease inhibitor (Ani s 6). It seems that allergens from *Anisakis* are stable to heat or cooking, and allergic reactions may be triggered by dead parasites in fish that have been well cooked [77]. It has been reported that these parasites can cause allergic sensitization among fish processing workers [90, 91]; therefore, possible allergic reactions to ingested fish could be directed to the contaminating parasite *Anisakis* and could be falsely diagnosed as fish allergy [89].

3.3.2 Tropomyosins and Other Shellfish Allergens

Tropomyosins are the major allergens responsible for ingestion-related allergic reactions because of crustaceans, while mollusks contain other less well-characterized allergens in addition to tropomyosin [92]. Interestingly, crustacean and mollusk allergens do not cross-react with fish allergens, and no reactivity between known allergens or homologous proteins has currently been demonstrated [89].

In the early 1980s, Hoffman et al. identified a heat-stable IgE reactive allergen in shrimps, which was later identified by Lehrer and colleagues in the brown shrimps as tropomyosin [93–95]. Shrimp tropomyosin (31.7 kDa, Met e 1) is a homodimer, which has an acidic isoelectric point. It is the major heat-stable shrimp allergen, and it has a highly conserved amino acid sequence among different invertebrate organisms, with up to eight IgE binding regions in shrimp, and is present in muscle and non-muscle cells [92].

Besides tropomyosin, other allergens have been identified and characterized in crustaceans. Five allergens have been registered from the black tiger shrimp (*Penaeus monodon*): tropomyosin Pen m 1 (38 kDa), arginine kinase as Pen m 2 (40 kDa) [96], myosin light chain 2 as Pen m 3 (19 kDa), sarcoplasmic calcium-binding protein as Pen m 4 (22 kDa) [97], and troponin C as Pen m 6 (17 kDa) [98]. The sarcoplasmic calcium-binding protein seems to be an important allergen particularly among the pediatric population. Tropomyosin is not only a crustacean allergen but has been confirmed in a number of mollusk species as well [99]. Tropomyosin has been demonstrated as one of the major allergens in squid, oysters, scallops, snails, and abalone [100]. Mollusks also contain other allergens, such as myosin heavy chain, hemocyanin, and amylase [99].

Tropomyosin seems to be the major allergen responsible for molecular and clinical cross-reactivity between crustaceans and mollusks, but also other inhaled invertebrates such as house dust mite and insects. While shellfish allergens do not cross-react with fish allergens, allergenic reactivity to *Anisakis*-contaminated fish might result from cross-reactivity to invertebrate tropomyosin [93].

3.3.3 Milk Allergens

Milk proteins are very heterogeneous with very few structural or functional common features. This heterogeneity is the consequence of their genetic polymorphism resulting in several variants for each protein. These variants are characterized by point substitutions of amino acids or by deletions of peptide fragments of varying size or by post-translational modifications such as phosphorylation and glycosylation. All of these modifications may affect the IgE-binding capacity and allergenicity [101]. Cow's milk proteins are classified as caseins or whey proteins. The casein fraction constitutes up to 80 % of the total protein and contains αs1-, αs2-, β-, and κ-caseins, as independent milk protein components, and three γ-caseins deriving from the hydrolysis of β-casein. γ1, γ2, and γ3 represent, respectively, the sequences 29–209, 106–209, and 108–209 of β-casein [102]. γ-Caseins are present in milk in minute quantities, while they are abundant in cheeses characterized by proteolytic ripening. Whey proteins are less abundant (20 % of the total protein) and Bos d 5 (β-lactoglobulin, BLG) is its main component (up to 50 % of whey proteins). Bos d 5 is a globular protein consisting of 162 amino acid residues with a molecular mass of 18.3 kDa. Its tertiary structure consists of nine antiparallel β-sheet structures forming a so-called β-barrel (or calyx) stabilized by formation of two disulfide bonds. It has one free thiol group which plays an essential role in the antioxidant activities of the protein. The interior of the calyx contains a hydrophobic pocket, allowing the binding of small hydrophobic molecules such as retinoids, fatty acids, vitamins, and cholesterol [103]. In solution, due to its very compact fold, Bos d 5 is highly stable to denaturation and resistant to proteolytic hydrolysis [25, 104].

Bos d 4 (α-lactalbumin, ALA) is a single-chain polypeptide of 123 amino acids corresponding to a molecular mass of 14,178 Da. It contains four disulfide bonds, and the primary structure is very similar to c-type lysozyme [105]. ALA normally binds one calcium ion (Kd $\sim 10^{-7}$ M); this binding dramatically changes the tertiary structure of the molecule from an open flexible form to a tight, compact globular structure, resulting in a major difference in size from 50 to 35 Å. Within the mammary gland, ALA serves as a regulator of the enzyme galactosyltransferase, which is responsible for the synthesis of lactose from galactose and glucose.

3.3.4   Egg Allergens

The four major allergens identified in hen's eggs (*Gallus gallus*) are ovomucoid (28 kDa, Gal d 1), ovalbumin (44 kDa, Gal d 2), ovotransferrin (78 kDa, Gal d 3), and lysozyme C (14 kDa, Gal d 4). These are also the most abundant proteins in egg white, representing 11, 54, 12, and 3.4 %, respectively, of the egg white proteins [106].

Ovomucoid, Gal d 1, consists of 186 amino acid residues and 25 % carbohydrate [107]. It is very stable under in vivo and in vitro conditions, being a serine protease inhibitor with nine disulfide bonds and no free –SH groups [108]. The carbohydrate chains are penta-antennary, heterogeneous, and partially sialylated [109, 110], resulting in substantial mass and charge heterogeneity of native ovomucoid. The carbohydrate chains seem to be unique and have not been reported to cause carbohydrate-based cross-reactivity [111].

Ovalbumin, Gal d 2, consists of 385 amino acid residues [112] and 3 % carbohydrate. It has one disulfide bond and four free –SH groups, which result in dimerization. Native ovalbumin displays considerable charge heterogeneity because of sequence variations, and phosphorylation in two sites with a reported ratio of 1:2:8 of zero, one, and two phosphate groups, respectively [113]. Finally, during storage in atmospheric air, ovalbumin rearranges to S-albumin, a conformationally different form, exposing an additional carboxylate group [114]. The post-translational modifications increase the sequence-derived molecular mass of 42,750 to 44,000–45,000 Da [111].

Ovotransferrin (also called conalbumin), Gal d 3, consists of 686 amino acids with 15 disulfide bonds, [115] and 3 % of carbohydrates. Charge heterogeneity arises from sequence variations and variations in bound Fe^{3+}. Ovotransferrin can bind two Fe^{3+} in association with binding of a bicarbonate anion [116], resulting in one extra negative charge per bound ferric ion. Ovotransferrin in egg white is normally without ferric ions. The theoretical pI of the dominating form is 6.69, and the molecular mass is 75,828 Da. The molecular mass of glycosylated ovotransferrin is approximately 77,000 Da [111].

Lysozyme, Gal d 4, consists of 129 amino acid residues, with four disulfide bonds and no free –SH groups [117]. Lysozyme has no post-translational modifications and is homogeneous with a theoretical pI of 9.3, and the molecular mass is 14,313 Da.

α-Livetin is, like chicken serum albumin (CSA), designated Gal d 5. CSA consists of 589 amino acid residues (Swiss-Prot entry P19121). It is homologous to mammalian serum albumins (47 and 44 % identity to human and bovine serum albumins, respectively). The protein has one potential glycosylation site and 35 cysteine residues. Based on similarity with other serum albumins, the –SH groups are expected to be linked in 17 disulfide bridges leaving one free –SH group giving rise to dimerization. The theoretical molecular mass of α-livetin is 66,815 Da and the theoretical pI 5.31 [111].

Gal d 6 is the yolk glycoprotein YGP42, a fragment of vitellogenin-1 (VTG-1). The VTG-derived proteins are the major yolk components; cleavage of VTG-1 and VTG-2 in the yolk produces apolipovitellins and phosvitins, which are components of the water-insoluble yolk granular lipoproteins. On the other hand, the C-terminal part of VTGs gives rise to yolk glycoproteins YGP40 and YGP42, which are major components of the yolk plasma [118].

Riboflavin-binding protein (RfBP) was reported as a minor IgE-reactive protein present in both egg white and yolk [119]. It is composed of 219 amino acid residues, containing carbohydrates that account for ~14 % of its molecular mass (30–35 kDa), phosphate moieties that contribute to its pI of about 4.0, and nine disulfide bonds which contribute to its thermal stability.

Chicken meat allergy is rare but there have been reports that it may be related to the bird-egg syndrome [120].

The low level of diversity among plant and animal food allergens appoint questions about the structural features that make these proteins allergenic. Features associated with allergenicity (presence of disulfide bonds, glycosylation, ligand binding, interactions with lipids, oligomerization) tend to enhance the structural integrity of the protein and confer resistance to gastrointestinal digestion, which allows an assumption that protein can reach the gut-associated lymphoid tissue (GALT).

Interaction with lipids can have the additional effect on the absorption by the gastrointestinal (GI) tract, and lipid binding could help to protect the allergen in the bloodstream and minimize elimination by proteolytic enzymes and by the components of innate immune system.

In conclusion, allergenic proteins are restricted to a small number of protein families. Numerous structural features of allergens contribute to the structural stability that enables them to resist digestion or heat inactivation. Understanding these structural features of food allergens can contribute to the design of novel strategies for the prevention of food allergy.

References

1. Hoffmann-Sommergruber K, Mills ENC. **Food allergen protein families and their structural characteristics and application in componet-resolved diagnosis: new data from the EuroPrevall project.** *Anal Bioanal Chem.* 2009, **395**(1):25–35.
2. Sampson HA: **Anaphylaxis and emergency treatment.** *Pediatrics* 2003, **111**(6 Pt 3):1601–1608.
3. Fleischer DM, Byrne AM, Malka-Rais J, Burks AW. (2010). **How do we know when peanut and tree nut allergy have resolved, and how do we keep it resolved?** *Clin. Exp. Allergy* 2010, **40**(9):1303–1311.
4. van Wijk F, Nierkens S, Hassing I, Feijen M, Koppelman SJ, de Jong GA, Pieters R, Knippels LM. **The effect of the food matrix on in vivo immune responses to purified peanut allergens.** *Toxicol Sci* 2005, **86**(2):333–41.
5. Van Wijk F, Nierkens S, de Jong W, Wehrens EJ, Boon L, van Kooten P, Knippels LM, Pieters R. **The CD28/CTLA-4-B7 signaling pathway is involved in both allergic sensitization and tolerance induction to orally administered peanut proteins.** *J Immunol* 2007, **178**(11):6894–900.
6. Schouten B, van Esch BC, Hofman GA, van den Elsen LW, Willemsen LE, Garssen J. **Acute allergic skin reactions and intestinal contractility changes in mice orally sensitized against casein or whey.** *Int Arch Allergy Immunol* 2008, **147**(2):125–34.
7. Hattori M, Nagasawa K, Ohgata K, Sone N, Fukuda A, Matsuda H, Takahashi K. **Reduced immunogenicity of beta-lactoglobulin by conjugation with carboxymethyl dextran.** *Bioconjugate Chem.* 2000, **11**(1):84–93.
8. Kobayashi K, Hirano A, Ohta A, Yoshida T, Takahashi K, Hattori M. **Reduced immunogenicity of beta-lactoglobulin by conjugation with carboxymethyl dextran differing in molecular weight.** *J Agric Food Chem.* 2001, **49**(2):823–831.
9. Kobayashi K, Yoshida T, Takahashi K, Hattori M. **Modulation of the T cell response to beta-lactoglobulin by conjugation with carboxymethyl dextran.** *Bioconjugate Chem.* 2003, **14**(1):168–176.
10. Nowak-Wegrzyn A, Bencharitiwong R, Schwarz J, David G, Eggleston P, Gergen PJ, Liu AH, Pongracic JA, Sarpong S, Sampson HA. **Mediator Release Assay for Assessment of Biological Potency of German Cockroach Allergen Extracts.** *J Allergy Clin Immunol.* 2009, **123**(4): 949–955.
11. Volgel L, Luttkopf D, Hatahet L, Haustein D, Vieths S. **Development of a functional in vitro assay as a novel tool for the standardization of allergen extracts in the human system.** *Allergy* 2005, **60**(8):1021–1028.
12. Mita H, Yasueda H, Akiyama K. **Affinity of IgE antibody to antigen influences allergen-induced histamine release.** *Clin. Exp. Allergy* 2000, **30**(11):1583–1589.
13. Torigoe C, Inman JK, Metzger H. **An unusual mechanism for ligand antagonism.** *Science* 1998, **281**(5376):568–572.

14. Ortega E, Schweitzer-Stenner R, Pecht I. **Possible orientational constrains determine secretory signals induced by aggregation of IgE receptors on mast cells.** *EMBO J* 1988, 7(13):4101–4109.

15. Johanson SG. **Anti-IgE antibodies in human serum.** *J Allergy Clin Immunol* 1986, 77(4):555–557.

16. Jarzab J, Gawlik R. **Immune complexes IgE/IgG in airborne allergy: increase during pollen season.** *J Investig Allergol Clin Immunol* 2000, 10(1):24–29.

17. Dibbern DA, Palmer GW, Williams PB, Bock SA, Dreskin SC. **RBL cells expressing human Fc epsilon RI are a sensitive tool for exploring functional IgE-allergen interactions: studies with sera from peanut-sensitive patients.** *J Immunol Methods* 2003, 274(1–2):37–45

18. Ebo DG. **Component-resolved allergy diagnosis: a new era?** *Verh K Acad Geneeskd Belg* 2011, 73(3–4):163–179

19. Valenta R, Lidholm J, Niederberger V, Hayek B, Kraft D, Grönlund H. **The recombinant allergen-based concept of component-resolved diagnostics and immunotherapy(CRD and CRIT).** *Clin Exp Allergy.* 1999, 29(7):896–904

20. Ott H, Baron JM, Heise R, Ocklenburg C, Stanzel S, Merk H-F, Niggemann B, Beyer K. **Clinical usefulness of microarray-based IgE detection in children with suspected food allergy.** *Allergy* 2008, 63(11):1521–1528

21. Breiteneder H, Radauer C: **A classification of plant food allergens.** *J Allergy Clin Immunol* 2004, 113(5):821–830; quiz 831

22. Jenkins JA, Breiteneder H, Mills EN: **Evolutionary distance from human homologs reflects allergenicity of animal food proteins.** *J Allergy Clin Immunol* 2007, 120(6):1399–1405

23. Radauer C, Breiteneder H: **Evolutionary biology of plant food allergens.** *J Allergy Clin Immunol* 2007, 120(3):518–525.

24. Kreis M, Forde BG, Rahman S, Miflin BJ, Shewry PR: **Molecular evolution of the seed storage proteins of barley, rye and wheat.** *J Mol Biol.* 1985, 183(3):499–502.

25. Maruyama N, Ichise K, Katsube T, Kishimoto T, Kawase S-i, Matsumura Y, Takeuchi Y, Sawada T, Utsumi S: **Identification of major wheat allergens by means of the *Escherichia coli* expression system.** *Eur J Biochem.* 1998, 255(3):739–745.

26. Breiteneder H, Mills EN. **Molecular properties of food allergens.** *J Allergy Clin Immunol* 2005, 115:14–23.

27. Robotham JM, Wang F, Seamon V, Teuber SS, Sathe SK, Sampson HA, et al. **Ana o 3, an important cashew nut (*Anacardium occidentale* L.) allergen of the 2S albumin family.** *J Allergy Clin Immunol* 2005, 115:1284–90.

28. Radauer C, Merima B, Wagner S, Mari A, Breiteneder H. **Allergens are distributed into few protein families and possess a restricted number of biochemical functions.** *J Allergy Clin Immunol* 2008, 121:847–852.

29. Dunwell JM, Khuri S, Gane PJ. **Microbial relatives of the seed storage proteins of higher plants: conservation of structure and diversification of function during evolution of the cupin superfamily.** *Microbiol Mol Biol Rev* 2000, 64:153–79.

30. Mills EN, Jenkins JA, Alcocer MJ, Shewry PR. **Structural, biological, and evolutionary relationships of plant food allergens sensitizing via the gastrointestinal tract.** *Crit Rev Food Sci Nutr* 2004, 44(5):379–407.

31. Dunwell JM, Purvis A, Khuri S. **Cupins: the most functionally diverse protein superfamily?** *Phytochemistry* 2004, 65:7–17.

32. Woo EJ, Marshall J, Bauly J, Chen JG, Venis M, Napier RM, Pickersgill RW. **Cristal structure of auxin-binding protein 1 in complex with auxin.** *EMBO J* 2002, 21(12):2877–2885.

33. Baumlein H, Braun H, Kakhovskaya IA, Shutov AD. **Seed storage proteins of spermatophytes share a common ancestor with desiccation proteins of fungi.** *J Mol Evol* 1995, 41:1070–1075.

34. Maruyama N, Adachi M, Takahashi K, Yagasaki K, Kohno M, Takenaka Y, Okuda E, Nakagawa S, Mikami B, Utsumi S. **Crystal structures of recombinant and native soybean beta-conglycinin beta homotrimer.** *Eur J Biochem* 2001, 268(12):3595–3604.

35. Adachi M, Kanamori J, Masuda T, Yagasaki K, Kitamura K, Mikami B, Utsumi S. **Crystal structure of soybean 11S globulin: glycin A3B4 homohexamer.** *Proc Natal Acad Sci USA* 2003, **100**(12):7395–7400.
36. Thompson MJ, Eisenberg D. **Transproteomic evidence of a loop-deletion mechanism for enhancing protein thermostability.** *J Mol Biol* 1999, **290**(2):595–604.
37. Polt G, Ahrazem O, Paschinger K, Ibanez MD, Salcedo G, Wilson IB. **Molecular and immunological characterization of the glycosylated orange allergen Cit s 1.** *Glycobiology* 2007, **17**(2):220–230.
38. van Loon LC, Rep M, Pieterse CM: **Significance of inducible defense-related proteins in infected plants.** *Annual Review of Phytopathology* 2006, **44**:135–162.
39. Breiteneder H, Ebner C: **Molecular and biochemical classification of plant-derived food allergens.** *J Allergy Clin Immunol* 2000, **106**(1 Pt 1):27–36.
40. Breiteneder H, Mills C: **Structural bioinformatic approaches to understand cross-reactivity.** *Mol Nutr Food Res* 2006, **50**(7):628–632.
41. Varjonen E, Vainio E, Kalimo K. **Antigliadin IgE—indicator of wheat allergy in atopic dermatitis.** *Allergy* 2000, **55**(4):386–391.
42. Takizawa T, Arakawa H, Tokuyama K, Morikawa A. **Identification of allergen fractions of wheat flour responsible for anaphylactic reactions to wheat products in infants and young children.** *Int Arch Allergy Immunol.* 2001, **125**(1):51–56.
43. Maruyama N, Ichise K, Katsube T, Kishimoto T, Kawase S-i, Matsumura Y, Takeuchi Y, Sawada T, Utsumi S: **Identification of major wheat allergens by means of the *Escherichia coli* expression system.** *Eur J Biochem* 1998, **255**(3):739–745.
44. Varjonen E, Vainio E, Kalimo K: **Life-threatening, recurrent anaphylaxis caused by allergy to gliadin and exercise.** *Clin Exp Allergy* 1997, **27**(2):162–166.
45. Palosuo K, Alenius H, Varjonen E, Koivuluhta M, Mikkola J, Keskinen H, Kalkkinen N, Reunala T: **A novel wheat gliadin as a cause of exercise-induced anaphylaxis.** *J Allergy Clin Immunol* 1999, **103**(5 Pt 1):912–917.
46. Varjonen E, Vainio E, Kalimo K, Juntunen-Backman K, Savolainen J: **Skin-prick test and RAST responses to cereals in children with atopic dermatitis. Characterization of IgE-binding components in wheat and oats by an immunoblotting method.** *Clin Exp Allergy* 1995, **25**(11):1100–1107.
47. Pastorello EA, Farioli L, Pravettoni V, Ortolani C, Ispano M, Monza M, Baroglio C, Scibola E, Ansaloni R, Incorvaia C *et al*: **The major allergen of peach (Prunus persica) is a lipid transfer protein.** *J Allergy Clin Immunol* 1999, **103**(3 Pt 1):520–526.
48. Pastorello EA, Ortolani C, Baroglio C, Pravettoni V, Ispano M, Giuffrida MG, Fortunato D, Farioli L, Monza M, Napolitano L *et al*: **Complete amino acid sequence determination of the major allergen of peach (Prunus persica) Pru p 1.** *Biol. Chem.* 1999, **380**(11):1315–1320.
49. Pastorello EA, Pravettoni V, Farioli L, Ispano M, Fortunato D, Monza M, Giuffrida MG, Rivolta F, Scibola E, Ansaloni R *et al*: **Clinical role of a lipid transfer protein that acts as a new apple-specific allergen.** *J Allergy Clin Immunol* 1999, **104**(5):1099–1106.
50. Sanchez-Monge R, Lombardero M, Garcia-Selles FJ, Barber D, Salcedo G: **Lipid-transfer proteins are relevant allergens in fruit allergy.** *J Allergy Clin Immunol* 1999, **103**(3 Pt 1):514–519.
51. Pastorello EA, Farioli L, Pravettoni V, Ispano M, Scibola E, Trambaioli C, Giuffrida MG, Ansaloni R, Godovac-Zimmermann J, Conti A, Fortunato D, Ortolani C. **The maize major allergen, which is responsible for food-induced allergic reactions, is a lipid transfer protein.** *J Allergy Clin Immunol* 2000, **106**(4):744–751.
52. Pastorello EA, D'Ambrosio FP, Pravettoni V, Farioli L, Giuffrida G, Monza M, Ansaloni R, Fortunato D, Scibola E, Rivolta F, Incorvaia C, Bengtsson A, Conti A, Ortolani C. **Evidence for a lipid transfer protein as the major allergen of apricot.** *J Allergy Clin Immunol* 2000, **105**(2 Pt 1):371–377.
53. Scheurer S, Pastorello EA, Wangorsch A, Kastner M, Haustein D, Vieths S. **Recombinant allergens Pru av 1 and Pru av 4 and a newly identified lipid transfer protein in the in vitro diagnosis of cherry allergy.** *J Allergy Clin Immunol* 2001, **107**(4):724–731.

54. James JM, Sixbey JP, Helm RM, Bannon GA, Burks AW. **Wheat alpha-amylase inhibitor: asecond route of allergic sensitization.** *J Allergy Clin Immunol* 1997, **99**(2):239–244.

55. Curioni A, Santucci B, Cristaudo A, Canistraci C, Pietravalle M, Simonato B, Giannattasio M. **Urticaria from beer: an immediate hypersensitivity reaction due to a 10-kDa protein derived from barley.** *Clin Exp Allergy* 1999, **29**(3):407–413.

56. Alvarez AM, Adachi T, Nakase M, Aoki N, Nakamura R, Matsuda T. **Classification of rice allergenic protein cDNAs belonging to the alpha-amylase/trypsin inhibitor gene family.** *Biochimica Biophysica Acta* 1995, **1251**(2):201–204.

57. Izumi H, Adachi T, Fujii N, Matsuda T, Nakamura R, Tanaka K, Urisu A, Kurosawa Y. **Nucleotide sequence of a cDNA clone encoding a major allergenic protein in rice seeds. Homology of the deduced amino acid sequence with members of alpha-amylase/trypsin inhibitor family.** *FEBS Letters* 1992, **302**(3):213–216.

58. Nakase M, Adachi T, Urisu A, Miyashita T, Alvarez AM, Nagasaka S, Aoki N, Nakamura R, Matsuda T: **Rice (*Oryza sativa* L.) α-Amylase Inhibitors of 14–16 kDa Are Potential Allergens and Products of a Multigene Family.** *J Agric Food Chem* 1996, **44**(9):2624–2628.

59. Teuber SS, Dandekar AM, Peterson WR, Sellers CL. **Cloning and sequencing of a gene encoding a 2S albumin seed storage protein precursor from English walnut (Juglans regia), a major food allergen.** *J Allergy Clin Immunol* 1998, **101**(6 Pt 1):807–814.

60. Menendez-Arias L, Moneo I, Dominguez J, Rodriguez R. **Primary structure of the major allergen of yellow mustard (Sinapis alba L.) seed, Sin a I.** *Eur. J Biochem* 1988, **177**(1):159–166.

61. Monsalve RI, Gonzalez de la Pena MA, Menendez-Arias L, Lopez-Otin C, Villalba M, Rodriguez R. **Characterization of a new oriental-mustard (Brassica juncea) allergen, Bra j IE: detection of an allergenic epitope.** *Biochem J* 1993, **293** (Pt 3):625–632.

62. Pastorello EA, Farioli L, Pravettoni V, Ispano M, Conti A, Ansaloni R, Rotondo F, Incorvaia C, Bengtsson A, Rivolta F, Trambaioli C, Previdi M, Ortolani C. **Sensitization to the major allergen of Brazil nut is correlated with the clinical expression of allergy.** *J Allergy Clin Immunol* 1998, **102**(6 Pt 1):1021–1027.

63. Burks AW, Williams LW, Connaughton C, Cockrell G, O'Brien TJ, Helm RM: **Identification and characterization of a second major peanut allergen, Ara h II, with use of the sera of patients with atopic dermatitis and positive peanut challenge.** *J Allergy Clin Immunol* 1992, **90**(6 Pt 1):962–969.

64. Kleber-Janke T, Crameri R, Appenzeller U, Schlaak M, Becker WM: **Selective cloning of peanut allergens, including profilin and 2S albumins, by phage display technology.** *Int Arch Allergy Immunol* 1999, **119**(4):265–274.

65. Stanley JS, King N, Burks AW, Huang SK, Sampson H, Cockrell G, Helm RM, West CM, Bannon GA: **Identification and mutational analysis of the immunodominant IgE binding epitopes of the major peanut allergen Ara h 2.** *Arch Biochem Biophys* 1997, **342**(2):244–253.

66. Pastorello EA, Varin E, Farioli L, Pravettoni V, Ortolani C, Trambaioli C, Fortunato D, Giuffrida MG, Rivolta F, Robino A *et al*: **The major allergen of sesame seeds (Sesamum indicum) is a 2S albumin.** *J Chromatography B Biomed Sci Appl* 2001, **756**(1–2):85–93.

67. Kelly JD, Hlywka JJ, Hefle SL: **Identification of sunflower seed IgE-binding proteins.** *Int Arch Allergy Immunol* 2000, **121**(1):19–24.

68. Burks AW, Williams LW, Helm RM, Connaughton C, Cockrell G, O'Brien T: **Identification of a major peanut allergen, Ara h I, in patients with atopic dermatitis and positive peanut challenges.** *J Allergy Clin Immunol* 1991, **88**(2):172–179.

69. Schuler MA, Ladin BF, Pollaco JC, Freyer G, Beachy RN: **Structural sequences are conserved in the genes coding for the alpha, alpha' and beta-subunits of the soybean 7S seed storage protein.** *Nucleic Acids Research* 1982, **10**(24):8245–8261.

70. Rabjohn P, Helm EM, Stanley JS, West CM, Sampson HA, Burks AW, Bannon GA: **Molecular cloning and epitope analysis of the peanut allergen Ara h 3.** *J Clin Investig* 1999, **103**(4):535–542.

71. Beardslee TA, Zeece MG, Sarath G, Markwell JP: **Soybean glycinin G1 acidic chain shares IgE epitopes with peanut allergen Ara h 3.** *Int Arch Allergy Immunol* 2000, **123**(4):299–307.

72. Nielsen NC, Dickinson CD, Cho TJ, Thanh VH, Scallon BJ, Fischer RL, Sims TL, Drews GN, Goldberg RB: **Characterization of the glycinin gene family in soybean.** *Plant Cell* 1989, **1**(3):313–328.
73. Pastorello EA, Conti A, Pravettoni V, Farioli L, Rivolta F, Ansaloni R, Ispano M, Incorvaia C, Giuffrida MG, Ortolani C: **Identification of actinidin as the major allergen of kiwi fruit.** *J Allergy Clin Immunol* 1998, **101**(4 Pt 1):531–537.
74. Ogawa T, Tsuji H, Bando N, Kitamura K, Zhu Y-L, Hirano H, Nishikawa K: **Identification of the soyabean allergenic proteins, Gly m Bd 30K, with the soybean seed 34 kDa oil-body-associated protein.** *Biosci. Biotechnol. Biochem* 1993, **57**:1030–1033.
75. Jenkins JA, Breiteneder H, Mills EN: **Evolutionary distance from human homologs reflects allergenicity of animal food proteins.** *J Allergy Clin Immunol* 2007, **120**(6):1399–1405
76. Kanduc D, Lucchese A, Mittelman A: **Individuation of monoclonal anti-HPV16 E7 antibody linear peptide epitope by computational biology.** *Peptides* 2001, **22**(12):1981–1985
77. Sharp MF, Lopata AL. **Fish allergy: in review.** *Clinic Rev Allerg Immunol* 2013, in press
78. Elsayed S, Apold J. **Immunochemical analysis of cod fish allergen M: locations of the immunoglobulin binding sites as demonstrated by the native and synthetic peptides.** *Allergy* 1983, **38**:449–459
79. Kretsinger RH. **Structure and evolution of calcium-modulated proteins.** *CRC Critical Reviews in Biochemistry*, 1980, **8**(2):119–174
80. Bugajska-Schretter A, Elfman L, Fuchs T, Kapiotis S, Rumpold H, Valenta R, Spitzauer S. **Parvalbumin, a cross-reactive fish allergen, contains IgE-binding epitopes sensitive to periodate treatment and Ca2+ depletion.** *J Allergy Clin Immunol* 1998, **101**:67–74.
81. Swoboda I, Bugajska-Schretter A, Verdino P, Keller W, Sperr WR, Valent P, Valenta R, Spitzauer S. **Recombinant carp parvalbumin, the major cross-reactive fish allergen: a tool for diagnosis and therapy of fish allergy.** *J Immunol* 2002, **168**:4576–4584.
82. Lim DL-C, Keng Hwee N, Fong Cheng Y, Kaw Yan C, Denise Li-Meng G, Lynette Pei-Chi S, Yoke Chin G, Hugo PSVB, Bee Wah L. **Parvalbumin—the major tropical fish allergen.** *Pediatric Allergy Immunol* 2008, **19**:399–407.
83. Van Do T, Hordvik I, Endresen C, Elssayed S. **The major allergen (parvalbumin) of cod-fish is encoded by at least two isotypic genes: cDNA cloning, expression and antibody binding of the recombinant allergens.** *Mol Immunol* 2003, **39**:595–602.
84. Perez-Gordo M, Lin J, Bardina L, Pastor-Vargas C, Cases B, Vivanco F, Cuesta-Herranz J, Sampson HA. **Epitope mapping of Atlantic salmon major allergen by peptide microarray immunoassay.** *Int Arch Allergy Immunol.* 2012, **157**(1):31–40
85. Das Dores S, Chopin C, Villaume C, Fleurence J, Gueant JL. **A new oligomeric parvalbumin allergen of Atlantic cod (Gad mI) encoded by a gene distinct from that of Gad cI.** *Allergy* 2002, **57**(72):79–83.
86. Rosmilah M, Shahnaz M, Masita A, Noormalin A, Jamaludin M. **Identification of major allergens of two species of local snappers: Lutjanus argentimaculatus (merah/red snapper) and Lutjanus johnii (jenahak/golden snapper).** *Trop Biomed* 2005, **22**(2):171–177
87. Tomm JM, van Do T, Jende C, Simon JC, Treudler R, von Bergen M, Averbeck M. **Identification of New Potential Allergens From Nile Perch (Lates niloticus) and Cod (Gadus morhua).** *J Investig Allergol Clin Immunol* 2013; **23**(3):159–167.
88. Lopata AL, Lehrer SB. **New insights into seafood allergy.** *Curr Opin Allergy Clin Immunol.* 2009, **9**(3):270–7.
89. Kirstein F, Horsnell WG, Nieuwenhuizen N, Ryffel B, Lopata AL, Brombacher F. **Anisakis-induced airway hyperresponsiveness is mediated by IFN-{gamma} in the absence of IL-4R{alpha}-responsiveness.** *Infect Immun.* 2010, **78**(9):4077–4086.
90. Nieuwenhuizen N, Lopata AL, Jeebhay MLF, Herbert DR, Robins TG, Brombacher F. **Exposure to the fish parasite Anisakis causes allergy airway hyperreactivity and dermatitis.** *J Allergy Clin Immunol* 2006, **117**(5):1098–1105
91. Lopata AL, O'Hehir RE, Lehrer SB. **Shellfish allergy.** *Clin Exp Allergy* 2010, **40**(6):850–858.
92. Hoffman DR, Day ED, Miller JS. **The major heat-stable allergen of shripm.** *Ann Allergy* 1981, **47**(1):17–22.

93. Daul J Slattery M, Reese G, Lehrer SB. **Identification of the major brown shrimp (Penaeus aztecus) allergen as the muscle protein tropomyosin.** *Int Arch Allergy Immunol* 1994, **105**(1):49–55.

94. Patrick SCL, Chu KH, Chow WK, Ansari A, Bandea CI, Kwan HS, Nagy SM, Gershwin EM. **Cloning, expression, and primary structure of Metapenaeus ensis tropomyosin, the major heat-stable shrimp allergen.** *J Allergy Clin Immunol* 1994, **94**(5):882–890.

95. Yu CJ, Lin YF, Chiang BL, Chow LP. **Proteomics and immunological analysis of a novel shrimp allergen, Pen m 2.** *J Immunol* 2003, **170**(1):445–453.

96. Taylor SL. **Molluscan shellfish allergy.** *Adv Food Nutr Res* 2008, **54**:139–177.

97. Mita H, Koketsu A, Ishizaki S, Shiomi K. **Molecular cloning and functional expression of allergenic sarcoplasmic calcium-binding proteins from Penaeus shrimps.** *J Sci Food Agric 2013*, **93**(7):1737–1742.

98. Abdel Rahman AM, Helleur RJ, Jeebhay MF, Lopata AL. **Characterization of Seafood Proteins Causing Allergic Diseases.** *Allergic Diseases* 2012, (5):107–140.

99. Lopata AL, Zinn C, Potter PC. **Characteristics of hypersensitivity reactions and identification of a unique 49 kD IgE-binding protein (Hal-m-1) in abalone (Haliotis midae).** *J Allergy Clin Immunol* 1997, **100**(5):642–648.

100. Blanc F, Bernard H, Alessandri S, Bublin M, Paty E, Leung SA, Patient KA, Wal J-M. **Update on optimized purification and characterization of natural milk allergens.** *Mol. Nutr. Food Res.* 2008, **52**(2):S166–S175.

101. Somma A, Ferranti P, Addeo F, Mauriello R, Chianese L. **Peptidomic approach based on combined capillary isoelectric focusing and mass spectrometry for the characterization of the plasmin primary products from bovine and water buffalo β-casein.** *J Chromatogr A* 2008, **1192**(2):294–300.

102. Kontopidis G, Holt C, Sawyer L. **The ligand-binding site of bovine beta-lactoglobulin: evidence for a function?** *J. Mol. Biol.* 2002, **318**(4):1043–1055

103. Stanic-Vucinic D, Stojadinovic M, Atanaskovic-Markovic M, Ognjenovic J, Gronlund H, van Hage M, Lantto R, Sancho AI, Cirkovic Velickovic T. **Structural changes and allergenic properties of beta-lactoglobulin upon exposure to high-intensity ultrasound.** *Mol. Nutr. Food Res.* 2012, **56**(12):1894–1905.

104. Lonnerdal B, Lien EL. **Nutritional and physiologic significance of α-lactalbumin in infants.** *Nutrition Reviews* 2003, **61**(9):295–305.

105. Osuga DT, Feeney RE. **Biochemistry of the egg-white proteins of the ratite group.** *Arch. Biochem. Biophy.* 1968,**124**:560–574.

106. Besler M, Steinhart H, Paschke A. **Allergenicity of hen's egg-white proteins: IgE-binding of native and deglycosylated ovomucoid.** *Food Agric. Immunol.* 1997, **9**(4):277–288.

107. Kato I, Schrode S, Kohr WJ, Laskowski M, Jr. **Chicken ovomucoid: Determination of its amino acid sequence, determination of the trypsin reactive site, and preparation of all three of its domains.** *Biochemistry* 1987, **26**(1):193–201.

108. Yamashita K, Kamerling JP, Kobata A. **Structural study of the carbohydrate moiety of hen ovomucoid—occurrence of a series of penta-antennary complex-type asparagine linked sugar chains.** *J. Biol. Chem.* 1982, **257**(21):12809–12814.

109. Yet MG, Chin CC, Wold F. **The covalent structure of individual N-linked glycopeptides from ovomucoid and asialofetuin.** *J. Biol. Chem.* 1988, **263**(1):111–7.

110. Jacobsen B, Hoffmann-Sommergruber K, Thordahl Have T, Foss N, Briza P, Oberhuber C, Radauer C, Alessandri S, Knulst AC, Fernandez-Rivas M, Barkholt V. **The panel of egg allergens, Gal d 1-Gal d 5: Their improved purification and characterization.** *Mol. Nutr. Food Res.* 2008, **52**(2):S176–S185.

111. Nisbet AD, Saundry RH, Moir AJG, Fothergill LA, Fothergill JE. **The complete amino acid sequence of hen ovalbumin.** *Eur. J. Biochem.* 1981, **115**(2):335–345.

112. Perlmann GE. **Enzymatic dephosphorylation of ovalbumin and plakalbumin.** *J Gen Physiol.* 1952, **35**(5):711–26.

113. Schafer A, Drewes W, Schwagele F. **Effect of storage temperature and time on egg white protein.** *Nahrung-Food* 1999, **43**(2):86–89.

114. Williams J, Elleman TC, Kingston IB, Wilkins AG, Kuhn KA. **The primary structure of hen ovotransferrin**. *Eur. J. Biochem.* 1982, **122**(2):297–303.
115. Aisen P, Leibman A, Reich HA. **Studies on binding of iron to transferrin and conalbumin**. *J. Biol. Chem.* 1966, **241**:1666–1671.
116. Canfield RE, Liu AK. The **disulfide bonds of egg white lysozyme (Muramidase)**. *J. Biol. Chem.* 1965, **240**:1997–2002.
117. Amo A, Rodriguez-Perez R, Blanco J, Villota J, Juste S, R, Moneo I, Caballero ML. **Gal d 6 is the second allergen characterized from egg yolk**. *J. Agric. Food Chem.* 2010, **58**(12):7453–7457.
118. Martos G, Pineda-Vadillo C, Miralles B, Alonso-Lebrero E, Lopez-Fandino R, Molina E, Belloque J. **Identification of an IgE reactive peptide in hen egg riboflavin binding protein subjected to simulated gastrointestinal digestion**. *J Agri Food Chem* 2012, **60**(20):5215–5220.
119. Sampson HA, Cooke SK. **Food allergy and the potential allergenicity-antigenicity of-microparticulated egg and cow's milk proteins**. *J. Am. Coll. Nutr.* 1990, **9**(4):410–417.
120. Bausela BA, García-Ara MC, Esteban MM, Martínez TB, Díaz Pena JM, Ojeda Casas JA. **Peculiarities of egg allergy in children with bird protein sensitization**. *Ann. Allergy Asthma Immunol.* 1997, **78**(2):213–216.
121. Farrell HM Jr, Jimenez-Flores R, Bleck GT, Brown EM, Butler JE, Creamer LK, Hicks CL, Hollar CM, Ng-Kwai-Hang KF. **Swaisgood HE. Nomenclature of the proteins of cows' milk—sixth revision**. *J Dairy Sci* 2004, **87**(6):1641–1674
122. Lisson M, Lochnit G, Erhardt G. **Genetic variants of bovine beta- and kappa-casein result in different immunoglobulin E-binding epitopes after in vitro gastrointestinal digestion**. *J Dairy Sci* 2013, **96**(9):5532–5543.

Chapter 4
Methods for Allergen Identification and Quantification in Food Matrices

Contents

Abbreviations

1D-PAGE	One-dimensional polyacrylamide gel electrophoresis
2D-PAGE	Two-dimensional polyacrylamide gel electrophoresis
ELISA	Enzyme-linked immunosorbent assay
ESI	Electrospray ionization
FEIA	Fluorenzymeimmunoassay
FRET	Fluorescence resonance energy transfer
LC	Liquid chromatography
LC–ESI–MS/MS	Liquid chromatography–electronspray ionization–tandem mass spectrometry
MALDI	Matrix-assisted laser desorption ionization
MS	Mass spectrometry
nsLTP	Nonspecific lipid transfer protein
PCR	Polymerase chain reaction
PMF	Peptide mass fingerprint
PNAs	Peptide nucleic acids
PVDF	Polyvinylidene fluoride
RAST	Radioallergosorbent test
SDS-PAGE	Sodium dodecyl sulfate-polyacrylamide gel electrophoresis
SPE	Solid-phase extraction
SPR	Surface plasmon resonance

T. Ćirković Veličković, M. Gavrović-Jankulović, *Food Allergens,*
Food Microbiology and Food Safety, DOI 10.1007/978-1-4939-0841-7_4,
© Springer Science+Business Media New York 2014

Summary The detection of allergenic ingredients in food products has received increased attention from the food industry and legislative and regulatory agencies over recent years. This has resulted in the improvement of applied safety measures which provide protection for food-allergic consumers. Several analytical approaches have been developed for the detection and quantification of allergens in food products. These methods target either the allergen itself or a corresponding allergen marker (peptide fragment or gene segment). The most popular methods for allergen detection at the protein level are antibody-based enzyme-linked immunosorbent assays (ELISA) and mass spectrometric methods. DNA-based methods include polymerase chain reaction (PCR), with real-time PCR enabling quantitative results regarding the potential presence of the culprit allergen. The employed analytical methods must be specific, highly sensitive, and should not be influenced by the presence of matrix components.

4.1   Principles of Protein Detection and Quantification Assays

Food allergies are estimated to affect about 2 % of the adult population in industrialized countries, and their prevalence in infants and children is about 6–8 % [1, 2]. More than 180 allergenic food proteins have been identified to date, with the major allergens occurring in common foods such as milk, egg, fish, crustaceans, peanut, soybean, wheat, and tree nuts [3]. Food allergens are proteins or glycoproteins with molecular mass of 5–70 kDa [1]; they usually represent the predominant protein fractions of a particular allergenic food and are resistant to proteolysis and stable during food processing [4].

The only effective management of food allergy is their complete avoidance from the diet. However, many of the allergenic foods are important nutrient sources and their complete exclusion from the diet is neither always possible nor desirable. Total avoidance is sometimes difficult for allergic individuals since processed food products contain a large variety of ingredients including allergenic food constituents. In sensitive patients, trace amounts of allergens can induce severe and even fatal reactions. As little as 30 μg of hazelnut is able to elicit an allergic reaction [5], and other predicted threshold values are 0.07 μg of milk, 0.003 μg of egg, 0.5 μg of peanut, and 0.3 mg of soybean [6]. Precise and sensitive methods for the detection and quantification of food allergens are essential to the food industry in order to guarantee the correct labeling of products and protection of allergic consumers. All substances purposely added to food products have to be labeled according to the European Food Labeling Directive. At the same time, in order to control allergens in foods, it is also important to know what quantity of an allergen can trigger an allergic reaction in an individual. However, the threshold at which all allergens can cause allergic reactions is not well known; therefore, it is not clear how sensitive the detection methods need to be [7]. Food products can be contaminated with foreign food constituents during shipping and storage, during processing, from carryover due to inadequate cleaning of shared processing equipment, or through rework of allergen-containing products [8]. In this regard, sensitive individuals may be inadvertently exposed to allergenic proteins by consumption of food products that are supposed to be free of a certain allergen [9].

The new EU Directive 2007/68/EC [10] provides a list of 14 groups of allergenic foods that manufacturers are required to declare on their labels if any of them are used as ingredients in prepacked foods, regardless of their quantity. The list includes gluten-containing cereals, crustaceans, eggs, fish, peanuts, soybeans, milk, nuts (almond, hazelnut, walnut, cashew, pecan nut, Brazil nut, pistachio nut, macadamia nut, and Queensland nut), celery, mustard, sesame seeds, lupine, mollusks, and sulfites. Reliable analytical methods for the detection of hidden allergens in foodstuff are required by the food industry and food control authorities to guarantee food safety for allergic consumers. To provide accurate and reliable information for allergic consumers, the food industry must develop reliable extraction and detection methods, meaning tailored and validated to the production and product characterization, as well as insight into possible contamination routes during processing and manufacturing [4].

In food allergen research, various chemical and physical methods are employed to detect proteins subsequently to their separation. Physical methods are mostly applied after chromatography. They are based on either spectroscopy, such as light absorption at certain wavelengths, or mass determination of peptides and their fragments with mass spectrometry (MS). Chemical methods are used after two-dimensional (2D) electrophoresis and employ staining with organic dyes (the most often used Coomassie Brilliant Blue), metal chelates, fluorescent dyes, complexing with silver, or prelabeling with fluorophores. Since all of these techniques are very different in terms of sensitivity, their usefulness for quantitative determinations varies significantly [11].

Several analytical approaches have been developed for the detection and quantification of allergens in food products [1, 8–10]. These can target either the allergen itself (one or several proteins) or an allergen marker (peptide fragment or gene segment). The most popular methods for allergen detection at the protein level are antibody-based enzyme-linked immunosorbent assays (ELISA), biosensors [12–16], and MS methods. DNA-based methods include polymerase chain reaction (PCR), with real-time PCR providing quantitative results on the potential presence of the culprit allergen. The employed analytical methods should be specific, highly sensitive, and should not be influenced by the presence of matrix components.

4.2 Antibody-Based Assays: Immunoassay and ELISA

Immunoassays are based on the specific interactions of antibody with antigen to provide quantitative information about antigen concentration in unknown samples. Historically, the most common technique used for allergen detection employed a radioactively labeled antigen or antibody and is known as radioimmunoassay. Originally, the term was reserved for techniques which involved competition for antibody binding between radiolabeled and unlabeled antigens. Alternative labels, such as enzymes (ELISA) or fluorochromes (fluorescence immunoassays), were quickly developed in place of radioisotopes. Further on, chemiluminescence technology has also been used in an attempt to increase sensitivity. Such immunoassays can be very sensitive and specific and therefore are commonly used for a great variety of measurements both in research and in analytical laboratories [17].

At the end of the 1960s, it became evident that the newly discovered immunoglobulin IgE was involved in allergy and that this immunoglobulin was present in the blood of allergic persons in higher concentrations than in nonallergic persons. There was a need for a simple in vitro method that would allow determination of "reaginic" antibody at nanogram-per-milliliter levels in serum of patients reactive to a given allergen. In this context, the paper disc radioallergosorbent test (RAST) method was developed for the assay of allergens. This method was used to diagnose allergy in vitro, to compare the "potency" of different allergen extracts, and to check the procedures for allergen extraction, storage, and further treatment [18]. Solid-phase radiobinding immunoassay or RAST is the simplest form of immunoassay, and has been useful for diagnosing specific IgE antibodies to the clinically suspected food allergen in the sera from allergic patients [19]. In these tests, antiserum is incubated with an antigen that has been immobilized by covalent attachment to agarose or polyacrylamide beads or by noncovalent "sticking" to plastic beads or, most commonly, to the wells of microtitre plates. Specific IgE is detected by incubation with radiolabeled antihuman IgE.

Competitive binding radioimmunoassay, or inhibition radioimmunoassay, combines high sensitivity and specificity with good reproducibility. In the classical version of this method, a fixed amount of radiolabeled antigen competes for a limited amount of specific antibody (sIgE) with unlabeled antigen. A standard curve is constructed using known amounts of unlabeled antigen which enables antigen concentration in unknown samples to be determined.

The market-leading RAST methodology was developed in the mid-1970s by Pharmacia Diagnostics AB, Uppsala, Sweden. In the late 1980s, Pharmacia Diagnostics AB replaced it with a superior test named the ImmunoCAP Specific IgE blood test, which is also described as CAP FEIA (fluorenzymeimmunoassay). The ImmunoCAP test is similar in concept to RAST but offers improved sensitivity. The main highlight of an ImmunoCAP test is the three-dimensional (3D) solid phase, which minimizes nonspecific binding by non-IgE binding antibodies. Reagent preparation is designed to reduce loss of conformational epitopes.

In 2010, the US National Institute of Allergy and Infectious Diseases recommended that the RAST measurements of specific IgE for the diagnosis of allergy be abandoned in favor of testing with more sensitive fluorescence enzyme-labeled assays [20].

A common method in protein biochemistry, sodium dodecyl sulfate-polyacrylamide gel electrophoresis (SDS-PAGE) is also frequently used to determine the presence or absence of protein pattern in food allergen extracts according to their molecular masses. The advantage of this analytical method is that it is inexpensive, and results can be obtained within hours; however, SDS-PAGE does not give any information on antigenicity/allergenicity of the analyzed protein extract. In addition, electrophoretic mobility of proteins can be affected by the presence of disulfide bonds [21].

Immunoblot methods for allergen analysis include Western blot and dot blot analysis. Western blot, as described by Towbin [22], requires proteins to be separated by molecular mass using PAGE. The proteins are then transferred to a membrane (polyvinylidene fluoride, PVDF, or nitrocellulose), and afterward subjected to antibody detection. A disadvantage of Western blotting is that proteins are most often tested in denatured form; so conformational epitopes may not be present in such partially unfolded structures. Another consequence of allergens being present in denatured form is that new IgE-binding epitopes that were hidden within the protein may be uncovered [23, 24]. It is also important to note that proteins are separated by molecular weight and those that are too large or too small for the resolution of the gel or blotting membrane may not be properly evaluated. The advantage of Western blotting is that protein bands can be individually analyzed to determine the changes in a specific allergen, and analysis is relatively fast, easy, and inexpensive. For dot blot analysis, the sample is directly adsorbed onto a membrane (i.e., nitrocellulose) and analyzed via antibody detection [25]. Dot blot, therefore, is a method of immunoblotting that does not involve denaturing conditions, and conformational epitopes are preserved. Because proteins are not separated by molecular weight, as in the case of Western blotting, the immunogenicity of the entire sample is analyzed. If the sample consists of an isolated protein, then single proteins may also be analyzed [26].

Enzyme-linked immunosorbent assay has been developed as a safer alternative to radioimmunoassays and had its breakthrough in biochemical and biomedical applications in the 1980s. Today, ELISA is the gold standard and most widely applied method for food allergen detection, offering a simple experimental design with suitable sensitivity for protein allergens in different sample matrices [27]. ELISA has become the method of choice for food producers and control agencies performing routine analysis of food allergen contaminations [4]. Besides its high sensitivity, the ELISA method is relatively cheap and easy to perform. Its high potential for standardization and automation enables high sample throughput which is essential for screening purposes [28]. Two formats of ELISA can be developed: competitive (direct) and sandwich. In the competitive ELISA, food allergen immobilized on the plate competes with the allergens from the sample to bind with the primary antibody labeled with enzyme. In the sandwich ELISA, two antibodies make a sandwich and the allergen is captured by these antibodies. The primary antibody immobilized on the solid phase captures the allergen, which is further detected by a secondary allergen-specific antibody labeled with enzyme or other marker molecule (i.e., biotin).

Practically, all ELISA formats have found their application in food analysis. Rapid ELISA test kits which provide qualitative and/or semiquantitative results within 30 and 60 min, respectively, have been developed and are commercially available for various foodstuff such as milk, soybean, peanut, hazelnut, almond, egg, crustaceans [reviewed in 29], and also for other food allergens [30, 31].

Immunological methods employ monoclonal and/or polyclonal antibodies that are raised in animals against purified allergens or other proteins specific for the allergenic commodity or, alternatively, human IgE. However, human IgE is characterized by a variability of sera obtained from individual patients which circumvent standardization and its limited availability prevents commercialization of methods based on human IgE. Some immunological methods are considered to be more laborious or more time consuming than others. ELISA and dipstick methods for the detection of food allergenic ingredients are commercially available and both methods utilize antibodies raised against allergenic commodities [32]. A multitude of ELISA methods have been developed for the detection of different food allergens [33–47] and numerous commercial test kits have become available during the last decade. Dipstick-based methods are less laborious and faster than ELISA, but they are not well suited for a quantitative assessment of the allergenic ingredient that they are designed to detect [28, 48].

Biosensors provide a rather novel approach for the detection of allergenic ingredients in food products [49]. In principle, they can be used to detect DNA or protein as analytes; however, they are currently mainly employed to detect protein analytes. This type of method benefits from short analysis times and has a potential for automation [28, 50]. An optical biosensor was used to develop both direct and sandwich immunoassays for the detection of proteins from various food samples (milk, egg, hazelnut, peanut, shellfish, and sesame). Affinity-purified polyclonal antibodies raised against the proteins were immobilized on the biosensor chip. Food samples were injected and the proteins that bound to the antibodies on the surface were detected by a shift in the resonance angle. By adding a second antibody in a sandwich assay, matrix effects could be overcome and the sensitivity and selectivity enhanced. Surface plasmon resonance (SPR)-based biosensors have been reported to be capable of detecting food allergens down to 1–12.5 µg/g in food samples and therefore achieve sensitivities comparable to immunoassays like ELISA [50].

Although accepted as standard methods of allergen measurement, the design of immunochemical-based methods varies with manufacturer. Antibodies can be raised to individual allergens or to total proteins [51]. Reference standards are usually represented by the raw unprocessed form of allergen [16]. While ELISA methods have been shown to be appropriate for the detection of low levels of protein allergens in complex matrices, discrepancies in quantitative results can arise due to limitations in protein extraction, lack of standard reference materials, variations in batch and cultivar sampling, and epitope modifications due to food processing [16, 52–55].

Because ELISAs could be affected by cross-reactivity and unpredictable effect of processing on food matrices and/or protein epitopes, positive ELISA results preferably require confirmatory analysis by nonimmunological techniques, such as PCR or MS, to corroborate data and to improve detection specificity [56].

4.3 MS-Based Approaches

In the past years, MS has become a highly employed methodology for protein analysis. The application of proteomic methodologies for the analysis of food allergens has been coined "allergenomics" [57]. Two aspects of employment of MS in allergenomics are foreseen: (1) elucidation of structural and functional features of food allergens and (2) development of a robust multi-allergen and quantitative method for trace analysis of food allergens [58]. MS allows proteins to be analyzed rapidly, accurately, and with high sensitivity, specificity, and reproducibility. Since their introduction in the late 1980s, matrix-assisted laser desorption ionization (MALDI) and electrospray ionization (ESI) have been the most widely employed soft ionization techniques for biomolecular analysis [59].

MS can be addressed either to measure the molecular mass of a protein/derived peptides or to determine additional structural details such as amino acid sequence and posttranslational modifications including process-induced modifications [59].

For the detection and identification of allergens, MS is frequently coupled with separation methods such as 1D- and 2D-PAGE, or liquid chromatography (LC). Two-dimensional electrophoresis has proven to be a reliable and efficient method for separation of a large number of proteins. Proteins are usually resolved according to their isoelectric point (pI) in isoelectric focusing (the first dimension) and subsequently to their molecular weight in SDS-PAGE (the second dimension). By 2D-PAGE, it is possible to separate simultaneously several thousand different protein spots. After visualization, protein spots excised from the gel are in situ digested with a protease (usually trypsin) and identified by MS or MS/MS. The first approach allows the protein identification by peptide mass fingerprint (PMF), by which the set of obtained masses is compared to the theoretically expected tryptic peptide masses for each entry in the database [59]. The second approach by MS/MS analysis (e.g., MALDI-TOF/TOF or ESI-MS/MS) provides structural information related to the amino acid sequence of detected peptides, making the search highly specific and discriminating [60].

Very often, proteins recognized by IgE from an allergic person's sera are identified and characterized using MS platforms. An 11S albumin was identified by PMF as a major hazelnut food allergen [61]. By using 2D maps of maize proteins, together with Western blot analysis with sera from patients allergic to maize and Orbitrap mass analysis, new allergens from this allergen source have been identified: vicilin, globulin-2, 50-kDa gamma-zein, endochitinase, thioredoxin, and trypsin inhibitor [62]. Combining fluorescent 2D differential gel electrophoresis with specific immunological detection as well as polypeptide sequencing by high-resolution MS, identification of two different isoforms of the allergen Ara h 1, the allergen Ara h 2, and six isoforms of the allergen Ara h 3/4 in 2D peanut protein maps was established [63].

A huge number of allergens have been identified and characterized from less common offending foods such as barley lipid transfer protein (LTP) [64], lettuce Lac s 1, a member of the nonspecific LTP (nsLTP) family [65], banana Mus a 5 [66], kiwifruit Act d 5 [67], peach Pru p 2 [68], etc.

An alternative to electrophoretic methods for complex protein extract separation is the liquid-phase separation (LC) coupled to MS analysis. The most effective LC/MS-based strategy is referred to as "shotgun" proteomics; in such an approach, the whole protein extract is trypsinized generating a complex mixture of peptides which is subsequently separated by LC prior to MS/MS.

The development and application of liquid chromatography–electrospray ionization–tandem mass spectrometry (LC–ESI–MS/MS)-based techniques for the investigation of allergens in food has considerably increased [3, 59, 69–71]. Various strategies can be applied for MS quantification of allergens and quantification, either at the protein level [72, 73] or at the peptide level [74, 75]. Mattarozzi et al. reported the development and validation of a shotgun proteomics LC–ESI–MS/MS-based method for the simultaneous detection and quantification of lupine allergens in biscuits and pasta in a single short run [76]. The method involves the use of a sample treatment incorporating solid-phase extraction (SPE) with size-exclusion columns for sample cleanup. The allergenic proteins β-conglutin, α-conglutin, γ-conglutin, and δ-conglutin were investigated by selecting and monitoring specific and unique target tryptic peptides. The method allowed rapid detection, unambiguous confirmation, and determination of lupine residues at trace levels in food products. In addition, MS is regarded as a complimentary tool for confirming litigated milk allergen ELISA results in industrial and regulatory settings [77]. Weber et al. employed the LC–MS/MS for detection of $\alpha s1$-casein in foods spiked with nonfat dry milk [78–80].

It is possible to detect trace amounts of allergen by identifying selected marker peptide. For example, peptide markers for peanuts were identified and used for the development of quantitative MS techniques [81–83].

MS methods are regarded as valuable tools for the detection and identification of traces of food allergens in different food matrices. However, limiting factors in their wider application are expensive equipment and highly trained personnel needed to operate such systems.

4.4   Matrix Effect and Food Processing in Detection of Food Allergens

Generally, an important issue related to detection of food allergens is that they are present usually in trace amounts and that their presence might be masked by the matrix. In antibody-based assays, the type of matrix present in the sample can influence antibody interactions with the analyte. The matrix can hinder the extraction of the analyte, or alternatively, co-extraction of matrix proteins alongside the analyte may occur, and these proteins can nonspecifically bind with antibodies, therefore giving false-positive results [4]. Hindered extractability because of interaction of the analyte with the matrix can also affect the detectability of food allergens by MS methods [84, 85]. As an example, interactions with matrix components, such as polyphenols and tannins from chocolate, impair the extractability of the analyte [86]. In addition, upon processing food allergens can be affected by denaturation with disruption of the tertiary and secondary structure, which can lead to modifications of the conformational epitopes. Conformational epitopes can also be modified through Maillard reactions or partial hydrolysis, contributing to protein aggregation and decreased solubility [4].

4.5   DNA-Based Methods for Allergen Detection

Methodologies based on genetics and molecular biology have become an interesting approach for tracking down the presence of trace amounts of allergens at any stage along the food supply chain [87, 88].

PCR has become a powerful technology for the rapid, sensitive, and specific detection of targeted DNA, although this technique represents an indirect investigation of the allergenic ingredients, since DNA may be considered a marker for protein allergen presence. The basic principle of the DNA-based methodology involves detection of a segment of the gene sequence coding for the allergen and amplifying only this DNA fragment to make it detectable [89]. A disadvantage of the DNA-based methods is that they do not detect the allergen itself and therefore serve as a surrogate for the allergen. To employ DNA as a surrogate marker, the particular allergenic food must have inherently high DNA content. Of the eight major allergenic foods and food groups, four are amenable to DNA-based detection: tree nuts, peanuts, fish, and crustacean shellfish. These four are inherently high in DNA and are likely to be present in foods as the whole plant or animal tissue, which contains both proteins and DNA. DNA-based detection is less appropriate for the remaining four allergenic foods: soy and wheat contain DNA, but are often present in food as protein fractions; eggs and milk contain inherently low level of DNA.

Advantages of DNA-based methods over protein-based methodologies are that the target DNA is efficiently extracted under harsh denaturing conditions and is less affected than proteins from food matrices. The DNA target sequence could be a gene coding for an allergenic protein or a species-specific region in the genome of the allergenic food [90, 91]. Although contaminants from food matrix may interfere with the PCR and lower the amplification efficiency, such obstacles are usually circumvented by diluting the template DNA and thereby concomitantly diluting the inhibitors [92].

The methodology of PCR is in vitro enzymatic synthesis of specific DNA sequences, using two oligonucleotide primers that hybridize to opposite strands and flank the region of interest in the target DNA. A repetitive series of cycles involving template denaturation, primer annealing, and the extension of the annealed primers by DNA polymerase results in the exponential accumulation of a specific fragment whose termini are defined by the 5′ ends of the primers. Because the primer extension products synthesized in one cycle can serve as a template in the next, the number of target DNA copies approximately doubles at every cycle. Thus, 20 cycles of PCR yields about millionfold (2^{20}) amplifications [93].

Classical PCR requires agarose gel electrophoresis for size-defined separation of the amplified PCR product and for visualization. Real-time PCR is a form of PCR where data are collected in real time as the reaction proceeds. In contrast to classical PCR, this method is based on fluorescence measurements to enable the visualization of PCR products during the amplified process. Real-time PCR uses either fluorescent DNA-binding dyes (e.g., SYBR Green) or fluorescent DNA probes, such as TaqMan® or hybridization probes based on fluorescence resonance energy transfer (FRET) technology that ensure the highest reliability of results.

PCR analysis allows the identification of traces of genomic DNA from the principal component and/or from contaminants in a food matrix [94]. The species-specific PCR amplification of the major hazelnut allergen Cor a 1 gene segment allowed the detection of 0.001 % (w/w) of hazelnut in commercial food products [95].

For simultaneous detection of hidden hazelnut and peanut traces in foodstuff, the duplex PCR method was developed, which detects specific traces down to 50 pg of the target DNA [96]. To overcome ambiguous interpretations on the low specificity of the primary sequences or of "carryover" contaminations that produce false positives, probes such as peptide nucleic acids (PNAs) that specifically hybridize target DNA have been developed. In such oligonucleotide analogs, the sugar-phosphate backbone has been replaced by a pseudopeptide chain of N-aminoethylglycine monomers contributing to a higher specificity. The duplex PCR method was a component of the PNA array device which allowed detecting simultaneously the presence of DNA from hazelnut and peanut, possible sources of hidden allergens in food products. The use of real-time PCR technique with a TagMan fluorescent probe for the detection of hazelnut in foods has been developed with a detection limit of 0.01 % (w/w) hazelnut in model pastry samples [97]. A real-time PCR targeting the major allergen gene of walnut allowed the detection of 0.24 ng of DNA and 0.01 % (w/w) walnut in model pastry samples [98].

In order to generate alternative detection methods for allergens for which effective protein-based assays were lacking, robust quantitative and sensitive methods for real-time PCR detection of celery, mustard (*Sinapsis alba* and *Brassica* sp.), and sesame in food were also developed [99].

Using a PCR–ELISA combination, Holzhauser et al. were able to detect very small quantities of hazelnut DNA in complex food matrixes [36]. In such a protocol, the PCR product is immobilized on a solid phase and after denaturation of the double stranded DNA, a sequence-specific fluorescein isothiocyanate (FITC)-labeled hybridization probe for DNA is added. Detection of the probe is performed with FITC–specific antibody–enzyme conjugate. With PCR-ELISA, less than 10 ppm of hazelnut was detected in complex food matrices.

However, as with protein-based assays, the food matrix was shown to influence results. It has been shown that the method of DNA extraction from food sample material can affect the PCR analysis due to the presence of inhibitors in the food matrices and quality of DNA molecules obtained, in terms of the length/fragmentation and quantity [100]. The food matrix and its physicochemical composition induce variability into the DNA extraction methods and in the efficacy of the DNA amplification [91, 101]. Thermal treatments such as roasting and autoclaving processes can reduce the extraction yield, and consequently detection of DNA by PCR methods [102]. Plant-derived foods produce secondary metabolites such as phenolic compounds, tannins, flavonoids, and alkaloids, whose presence in the extracted DNA material can interfere with the analysis and inhibit the PCR [103]. The fat material and polyphenol compounds of nuts can act as polymerase inhibitors [104].

Compared to ELISA, PCR methods for allergen detection have the advantage of rapid test development within 7–10 days when the referring DNA sequence is known [105]. In addition, DNA represents a more stable target molecule in processed food samples than a distinct protein. The genotype is independent of climatic, seasonal, and local influences. Protein-based methods are directed against an epitope/s, particular regions of a protein. These protein segments can be denatured by processing the food and as a consequence cannot be recognized by the detection antibody, although the allergenic activity may be still intact. Another striking advantage of real-time PCR is better sensitivity, caused by the use of three specific detection molecules (two primers and a probe).

Real-time PCR is a very sensitive technique detecting down to a few molecules of DNA. However, the sensitivity of this assay is highly dependent on the sample preparation. This may vary from food product to food product. Sensitivities in the range of 5–50 mg kg^{-1} in heterogeneous matrices are generally achieved by using sample quantities between 500 mg and 2 g and by using a DNA extraction and purification method, which yields DNA pure enough for the subsequent PCR analysis. Increased sensitivity of a PCR-based kit for soy, with a sensitivity of 1.5 mg kg^{-1}, has been developed, and it was shown to have an approximately 500 times lower detection limit than the one obtained for ELISA [106].

DNA-based methods are not well suited for the detection of allergenic commodities characterized by high protein content and a low DNA content (e.g., eggs). In contrast, DNA-based methods may be a good choice for allergen commodities where protein content is low (e.g., celery).

Over the past several decades, food allergies have become an important food-safety issue and development of analytical methods for the detection and quantification of allergenic foods has become important for the protection of food-allergic consumers. Analytical methods used for the detection of residues of allergenic foods must not only be highly specific, sensitive, and rugged enough to be applicable in food matrices but also reliable. Because of the diverse nature of food allergens and respective food matrices, the choice of detection methods and achievement of reliable results can be quite challenging.

References

1. Pooms RE, Klein CL, Anklam E: **Methods for allergen analysis in food: a review**. *Food Addit. Contam*. 2004, **21**(1):1–31.
2. Mills ENC, Mackie AR, Burney P, Beyer K, Frewer L, Madsen C, Botjes E, Crevel RWR, van Ree R. **The prevalence, cost and basis of food allergy across Europe**. *Allergy* 2007, **62**(7):717–722.
3. Monaci L, Visconti A: **Mass spectrometry/based proteomics methods for analysis of food allergens**. *TrAC-Trends Anal. Chem*. 2009, **28**(5):581–591.
4   Cucu T, Jacxsens L, De Meulenaer B: **Analysis to support allergen risk management: which way to go**? *J Agri. Food Chem*. 2013, **61**(3):5624–5633.
5. Cochrane S, Salt LJ, Wantling E, Rogers A, Coutts J, Ballmer-Weber BK, Fritsche P, Fernandez MI, Reig I, Knults A, Le TM, Asero R, Beyer K, Golding M, Crevel M, Mills ENC, Mackie AR: **Development of a standardized low-dose double-blind placebo-controlled challenge vehicle for the EuroPrevall project**. *Allergy* 2012, **67**(1):17–113.
6. Bindslev-Jensen C, Briggd D, Osterballe M: **Can we determine a threshold level for allergenic foods by stastical analysis of published data in the literature**? *Allergy* 2002, **57**(8):741–746.

7. A. Barrosa A, Cosme F. **Allergenic Proteins in Food.** *Food Technol. Biotechnol.* 2013, **51**(2):153–158.
8. Huggett AC, Hischenhuber C. **Food manufacturing initiatives to protect the allergic consumer.** *Allergy* 1998, **53**(46), 89–92.
9. Poms RE, Capelletti C, Anklam E. **Effect of roasting history and buffer composition on peanut protein extraction efficiency.** *Mol. Nutr. Food Res.* 2004, **48**(6):459–464.
10. European Parliament and Council. **Directive 2007/68/EC.** *Official J Eur Union* 2007;L310: 11–14.
11. Westermeier R, Marouga R. **Protein Detection Methods in Proteomics Research.** *Bioscience Reports* 2005, **25**(1/2):19–32.
12. Bremer MGEG, Smits NGE, Haasnoot W: **Biosensor immunoassays for traces of hazelnut protein in olive oil.** *Anal. Bioanal. Chem.* 2009, **395**(1):119–126.
13. Maier I, Morgan MRA, Lindner W, Pittner F: **Optical resonance-enhanced absorption-based near-field immunochip biosensor for allergen detection.** *Anal. Bioanal. Chem.* 2008, **80**(8):2694–2703.
14. Pollet J, Delport F, Janssen KPF, Tran DT, Wouters J, Verbiest T, Lammertyn J: **Fast and accurate peanut allergen detection with nanobead enhanced optical fiber SPR biosensor.** *Talanta* 2011, **83**(5):1436–1441.
15. Trashin S, Cucu T, Devreese B, Adriaens A, De Meulenaer B: **Development of a highly sensitive and robust Cor a 9 specific enzyme-linked immunosorbent assay for the detection of hazelnut traces.** *Anal. Chim. Acta* 2011, **708**(1–2):116–122.
16. Yman IM, Eriksson A, Johansson MA, Hellenas KE: **Food allergen detection with biosensor immunoassay.** *J. AOAC Int.* 2006, **89**(3):856–861.
17. Johnstone A, Thorpe R. **Immunochemistry in practice.** Blackwell Scientific Publications, Oxford 1987.2nd Ed, pp 241.
18. Ceska M, Eriksson R, Varga JM. **Radioimmunosorbent assay of allergens.** *J Allergy Clin Immunol* 1972, **49**(1):1–9.
19. Hoffman DR, Haddad ZH. **Diagnosis of IgE mediated hypersensitivity reaction to foods by radioimmunoassay.** *J Allergy Clin Immunol* 1974, **54**(3):164–173.
20. NIAID-Sponsored Expert Panel. **Guidelines for the Diagnosis and Management of Food Allergy in the United States: Report of the NIAID-Sponsored Expert Panel.** *J Allergy Clin Immunol* 2010, **126**(6):S1–S58.
21 Grozdanović MM, Drakulić BJ, Gavrović-Jankulović M. **Conformational mobility of active and E-64-inhibited actinidin.** *Biochim Biophys Acta.* 2013, **1830**(10):4790–4799.
22. Towbin H, Staehelin T, Gordon J. **Electrophoretic transfer of proteins from polyacrylamide gels to nitrocellulose sheets: procedure and some applications.** *Proc Natl Acad Sci USA* 1979, **76**(9):4350–4354.
23. Grozdanovic M, Popovic M, Polovic N, Burazer L, Vuckovic O, Atanaskovic-Markovic M, Lindner B, Petersen A, Gavrovic-Jankulovic M. **Evaluation of IgE reactivity of active and thermally inactivated actinidin, a biomarker of kiwifruit allergy.** *Food Chem Toxicol.* 2012, **50**(3–4):1013–8.
24. Herndl A, Marzban G, Kolarich D, Hahn R, Boscia D, Hemmer W, Maghuly F, Stoyanova E, Katinger H, Laimer M. **Mapping of Malus domestica allergens by 2-D electrophoresis and IgE-reactivity.** Electrophoresis 2007, 28(3):437–448.
25. Singh M, Knox R. **Grass pollen allergens: antigenic relationships detected using monoclonal antibodies and dot blotting immunoassay.** *Int Arch Allergy Immunol* 1985, **78**(3):300–304.
26. Shriver SK, Yang WW. **Thermal and Nonthermal Methods for Food Allergen Control.** *Food Eng. Rev.* 2011, 3(1):26–43.
27. Hebling CH, McFarland MA, Callahan JH, Ross MM: **Global proteomic screening of protein allergens and advanced glycation endproducts in thermally processed peanuts.** *J. Agric. Food Cehm.* 2013, **61**(24):5638–5648.
28. Goldsby RA, Kindt TJ, Osborne BA, Kuby J. **Enzymelinked immunosorbent assay.** In Immunology. New York: W.H. Freeman, & Company 2003, pp 148–150.

29. van Hengel AJ. **Food allergen detection methods and the challenge to protect food-allergic consumers.** *Anal Bioanal Chem* 2007, **389**(1):111–118.
30. Gavrovic-Jankulovic M., Spasic M, Cirkovic Velickovic T, Stojanovic M, Inic-Kanada A, Dimitrijevic L, Lindner B, Petersen A, Becker W-M, Jankov RM. **Quantification of the thaumatin-like kiwi allergen by a monoclonal antibody-based ELISA.** *Mol Nutr Food Res* 2008, **52** (6):701–707.
31. Abedini S, Sankian M, Falak R, Tehrani M, Talebi F, Shirazi FG, Varasteh A-R. **An approach for detection and quantification of fruits' natural profilin: Natural melon profilin as a model.** *Food Agric Immunol* 2011, **22**(1):47–55.
32. Stephan O, Vieths S. **Development of a Real-Time PCR and a Sandwich ELISA for Detection of Potentially Allergenic Trace Amounts of Peanut (Arachis hypogaea) in Processed Foods.** *J Agric Food Chem* 2004, **52**(12):3754–3760.
33. Hefle SL, Bush RK, Yunginger JW, Chu SF. **A sandwich enzyme-linked immunosorbent assay (ELISA) for the quantification of selected peanut proteins in foods.** *J Food Protection* 57:419–423.
34. Duffort OA, Polo F, Lombardero M, Díaz-Perales A, Sánchez-Monge R, García-Casado G, Salcedo G, Barber D. **Immunoassay To Quantify the Major Peach Allergen Pru p 3 in Foodstuffs. Differential Allergen Release and Stability under Physiological Conditions.** *J. Agric. Food Chem.*, 2002, **50**(26), pp 7738–7741.
35. Scharf A, Kasel U, Wichmann G, Besler M. **Performance of ELISA and PCR Methods for the Determination of Allergens in Food: An Evaluation of Six Years of Proficiency Testing for Soy (Glycine max L.) and Wheat Gluten (Triticum aestivum L.)** *J. Agric. Food Chem.*, 2013, **61**(43), pp 10261–10272.
36. Holzhauser T, Stephan O, Vieths S. **Detection of Potentially Allergenic Hazelnut (Corylus avellana) Residues in Food: A Comparative Study with DNA PCR-ELISA and Protein Sandwich-ELISA.** *J. Agric. Food Chem.*, 2002, **50**(21), pp 5808–5815.
37. Werner MT, Fæste CK, Egaas E. **Quantitative Sandwich ELISA for the Determination of Tropomyosin from Crustaceans in Foods.** *J. Agric. Food Chem.*, 2007, **55**(20), pp 8025–8032.
38. Pelaez-Lorenzo C, Diez-Masa JC, Vasallo I, de Frutos M. **Development of an Optimized ELISA and a Sample Preparation Method for the Detection of β-Lactoglobulin Traces in Baby Foods.** *J. Agric. Food Chem.*, 2010, **58**(3), pp 1664–1671.
39. Schneider N, Weigel I, Werkmeister K, Pischetsrieder M. **Development and Validation of an Enzyme-Linked Immunosorbent Assay (ELISA) for Quantification of Lysozyme in Cheese.** *J. Agric. Food Chem.*, 2010, **58**(1), pp 76–81.
40. Patrick W, Hans S, Angelika P. **Determination of the Bovine Food Allergen Casein in White Wines by Quantitative Indirect ELISA, SDS–PAGE, Western Blot and Immunostaining.** *J. Agric. Food Chem.*, 2009, **57**(18), pp 8399–8405.
41. Gruber P, Becker W-M, Hofmann T. **Influence of the Maillard Reaction on the Allergenicity of rAra h 2, a Recombinant Major Allergen from Peanut (Arachis hypogaea), Its Major Epitopes, and Peanut Agglutinin.** *J. Agric. Food Chem.*, 2005, **53**(6), pp 2289–2296.
42. Sancho AI, Foxall R, Browne T, Dey R, Zuidmeer L, Marzban G, Waldron KW, van Ree R, Hoffmann-Sommergruber K, Laimer M, Mills ENC. **Effect of Postharvest Storage on the Expression of the Apple Allergen Mal d 1.** *J. Agric. Food Chem.*, 2006, **54**(16), pp 5917–5923.
43. Holden L, Fste CK, Egaas E. **Quantitative Sandwich ELISA for the Determination of Lupine (Lupinus spp.) in Foods.** *J. Agric. Food Chem.*, 2005, **53**(15), pp 5866–5871.
44. Lifrani A, Santos JD, Dubarry M, Rautureau M, Blachier F, Tome D. **Development of Animal Models and Sandwich-ELISA Tests to Detect the Allergenicity and Antigenicity of Fining Agent Residues in Wines.** *J. Agric. Food Chem.*, 2009, **57**(2), pp 525–534.
45. Holden L, Haugland Moen L, Sletten GBG, Dooper MMBW. **Novel Polyclonal–Monoclonal-Based ELISA Utilized To Examine Lupine (Lupinus Species) Content in Food Products.** *J. Agric. Food Chem.*, 2007, **55**(7), pp 2536–2542.

46. Sancho AI, Foxall R, Rigby NM, Browne T, Zuidmeer L, van Ree R, Waldron KW, Mills ENC. **Maturity and Storage Influence on the Apple (Malus domestica) Allergen Mal d 3, a Nonspecific Lipid Transfer Protein.** *J. Agric. Food Chem.*, 2006, **54** (14), pp 5098–5104.

47. Shimizu Y, Nakamura A, Kishimura H, Hara A, Watanabe K, Saeki H. **Major Allergen and Its IgE Cross-Reactivity among Salmonid Fish Roe Allergy.** *J. Agric. Food Chem.*, 2009, **57**(6), pp 2314–2319.

48. Schubert-Ullrich P, Rudolf J, Ansari P, Galler B, Fuhrer M, Molinelli A, Baumgartner S. **Commercialization rapid immunoanalytical tests for determination of allergenic food proteins: an overview.** *Anal Bioanal Chem* 2009, **395**(1):69–81.

49. Haasnoot W, Smits N, Kemmers-Voncken A, Bremer M. **Fast biosensor immunoassays for the detection of cows' milk in the milk of ewes and goats.** *J Dairy Res.* 2004, **71**:322–329.

50. Yman IM, Eriksson A, Johansson MA, Hellenäs KE. **Food allergen detection with biosensor immunoassays.** *J AOAC Int* 2006, **89**(3):856–861.

51. Goodwin PR: **Food allergen detection methods: a coordinated approach.** *J. AOAC Int.* 2004, **87**(6):1383–1390.

52. Whitaker TB, Williams KM, Trucksess MW, Slate AB: **Immunochemical analytical methods for the determination of peanut proteins in foods.** *J. AOAC Int.* 2005, **88**(1):161–174.

53. Hurst WJ, Krout ER, Burks WR: **A comparison of commercially available peanut ELISA test kits on the analysis of samples of dark and milk chocolate.** *J. Immunoassay Immunochem.* 2002, **23**(4):451–459.

54. Poms RE, Capelletti C, Anklam E. **Effect of roasting history and buffer composition on peanut protein extraction efficiency.** *Mol. Nutr. Food Res.* 2004, **48**(6):459–464.

55. Scaravelli E, Brohee M, Marchelli R, Van Hengel AJ. **The effect of heat treatment on the detection of peanut allergens as determined by ELISA and real-time PCR.** *Anal Bioanal Chem* 2009, **395**(1):127–137

56. Holden L, Moen LH, Sletten GB, Dooper MM. **Novel polyclonal-monoclonal-based ELISA utilized to examine lupine (*Lupinus* species) content in food products.** *J. Agric. Food Chem.* 2007, **55**(7):2536–2542.

57. Akagawa M, Handoyo T, Ishii T, Kumazawa S, Morita N, Suyama K. **Proteomic analysis of wheat flour allergens.** *J. Agric. Food Chem.* 2007, **55**(17):6863–6870.

58. Pedreschi R, Nørgaard J, Maquet A. **Current Challenges in Detecting Food Allergens by Shotgun and Targeted Proteomic Approaches: A Case Study on Traces of Peanut Allergens in Baked Cookies.** *Nutrients* 2012, **4**(2): 132–150.

59. Picariello G, Mamone G, Addeo F, Ferranti P. **The frontiers of mass spectrometry-based techniques in food allergenomics.** *J. Chromatogr. A.* 2011, **1218**(42):7386–7398.

60. Mann M, Hendrickson RC, Pandey A. **Analysis of proteins and proteome by mass spectrometry.** *Annu. Rev. Biochem.* 2001, **70**:437–473.

61. Beyer K, Grishina L, Bardina A, Grishin HA, Sampson HA. **Identification of an 11S globulin as a major hazelnut food allergen in hazelnut-induced systemic reactions.** *J. Allergy Clin. Immunol.* 2002, **110**(3):517–523.

62. Fasoli E, Pastorello EA, Farioli L, Scibilia J, Aldini G, Carini M, Marocco A, Boschetti E, Righetti PG. **Searching for allergens in maize kernels via proteomic tools.** *J. Proteomics* 2009, **72**(3):501–510.

63. Chassaigne H, Treỉgoat V, Nørgaard JV, Maleki SJ, van Hengel AJ. **Resolution and identification of major peanut allergens using a combination of fluorescence two-dimensional differential gel electrophoresis, Western blotting and Q-TOF mass spectrometry.** *J. Proteomics* 2009, **72**(3):511–526.

64. Garcia-Casado G, Crespo JF, Rodriguez J, Salcedo G. **Isolation and characterization of barley lipid transfer protein and protein Z as beer allergens.** *J. Allergy Clin. Immunol.* 2001, **108**(4):647–649.

65. Hartz C, San Miguel-Moncín Mdel M, Cisteró-Bahíma A, Fötisch K, Metzner KJ, Fortunato D, Lidholm J, Vieths S, Scheurer S. **Molecular characterisation of Lac s 1, the major allergen from lettuce (*Lactuca sativa*).** *Mol. Immunol.* 2007, **44**(11):2820–2830.

66. Aleksic I, Popovic M, Dimitrijevic R, Andjelkovic U, Vassilopoulou E, Sinaniotis A, Atanas-kovic-Markovic M, Lindner B, Petersen A, Papadopoulos NG, Gavrovic-Jankulovic M. **Molecular and immunological characterization of Mus a 5 allergen from banana fruit.** *Mol Nutr Food Res.* 2012, **56**(3):446–53.
67. Tuppo L, Giangrieco I, Palazzo P, Bernardi ML, Scala E, Carratore V, Tamburrini M, Mari A, Ciardiello MA. **Kiwellin, a Modular Protein from Green and Gold Kiwi Fruits: Evidence of in Vivo and in Vitro Processing and IgE Binding.** *J. Agric. Food Chem.* 2008, **56**(10): 3812–3817.
68. Palacín A, Tordesillas L, Gamboa P, Sanchez-Monge R, Cuesta-Herranz J, Sanz M L, Barber D, Salcedo G, Díaz-Perales A. **Characterization of peach thaumatin-like proteins and their identification as major peach allergens.** *Clin. Exp. Allergy* 2010, **40**(9):1422–1430.
69. Mari A, Ciardiello MA, Tamburrini M, Rasi C, Palazzo P. **Proteomic analysis in the identification of allergenic moleculs.** *Expert Rev Proteomics* 2010, **7**(5):723–734.
70. Sancho AI, Mills EN. **Proteomic approach for qualitative and quantitative characterization of food allergens.** *Regul Toxicol Pharmacol.* 2010, **58**(3):S42–S46
71. Faeste CK, Ronning HT, Christians U, Granum PE. **Liquid chromatography and mass spectrometry in food allergen detection.** *J Food Prot.* 2011, **74**(2):316–345.
72. Czerwenka C, Maier I, Potocnik N, Pittner F, Lindner W. **Absolute quantification of beta-lactoglobulin by protein liquid chromatography-mass spectrometry and its application to different milk products.** *Anal. Chem.* 2007, **79**(14);5165–5172.
73. Monaci L, van Hengel AJ. **Development of a method for the quantification of whey allergen traces in mixed-fruit juices based on liquid chromatography with mass spectrometry detection.** *J Chromatogr A.* 2008, **1192**(1):113–120.
74. Careri M, Costa A, Elviri L, Lagos J, Mangia A, Terenghi M, Cereti A, Garoffo L. **Use of specific peptide biomarkers for quantitative confirmation of hidden allergenic peanut proteins Ara h 2 and Ara Ill ¾ for food control by liquid chromatography-tandem mass spectrometry.** *Anal Bioanal Chem.* 2007, **389**(6):1901–1907.
75. Bignardi C, Elviri L, Penna A, Careri M, Mangia A. **Particle-packed column versus silica-based monolithic column for liquid chromatography-electrospray-linear ion trap-tandem mass spectrometry multiallergen trace analysis in foods.** *J Chromatogr A.* 2010, **1217**(48):7579–85.
76. Mattarozzi M, Bignardi C, Elviri L, Careri M. **Rapid shotgun proteomics liquid chromatography-electrospray ionization-tandem mass spectrometry-based method for the lupin (Luoinus albus L.) multi-allergen determination in foods.** *J. Agric. Food Chem.* 2012, **60**(23):5841–5846.
77. Newsome GA, Scholl PF. **Quantification of allergenic bovine milk αs1-casein in baked goods using an intact** [15]**N-labeled protein internal standard.** *J. Agric. Food Chem.* 2013, **61**(24):5659–5668.
78. Weber D, Raymond P, Ben-Rejeb S, Lau B: **Development of a liquid chromatography-tandem mass spectrometry method using capillary liquid chromatography and nano-electrospray ionization-quardipole time-of-flight hybrid mass spectrometer for the detection of milk allergens.** *J. Agric. Food Chem.* 2006, **54**(5):1604–1610.
79. Heick J, Fischer M, Popping B: **First screening method for the simultaneous detection of seven allergens by liquid chromatography mass spectrometry.** J Chromatogr A. 2011, **1218**(7):938–43.
80. Heick J, Fischer M, Kerbach S, Tamm U, Popping B. **Application of liquid chromatography tandem mass spectrometry method for the simultaneous detection of seven allergenic foods in flour and bread and comparison of the method with commercially available ELISA rest kits.** *J. AOAC Int.* 2011, **94**(4):1060–1068.
81. Apostolovic D, Luykx D, Warmenhoven H, Verbart D, Stanic-Vucinic D, de Jong GA, Velickovic TC, Koppelman SJ. **Reduction and alkylation of peanut allergen isoforms Ara h 2 and Ara h 6; characterization of intermediate- and end products.** *Biochim Biophys Acta.* 2013, **1834**(12):2832–2842

82. Chassaigne H, Norgaard JV, van Hengel AJ: **Proteomics-based approach to detect and identify major allergens in processed peanuts by capillary LC-Q-TOF (MS/MS).** *J. Agric. Food Chem.* 2007, **55**(11):4461–4473.

83. Shefcheck KJ, Musser SM: **Conformation of the allergenic peanut protein, Ara h 1, in a model food matrix using liquid chromatography/tandem mass spectrometry (LC-MS/MS).** *J. Agric. Food Chem.* 2006, **54**(21):7953–7959.

84. Lutter P, Parisod V, Weymuth H. **Development and validation of a method for the quantification of milk proteins in food products based on liquid chromatography with mass spectrometric detection.** *J. AOAC Int.* 2011, **94**(4):1043–1059.

85. Monaci L, Losito I, Palmisano F, Visconti A. **Identification of allergenic milk proteins markers in fined white wines by capillary liquid chromatography-electrospray ionization-tandem mass spectrometry.** *J. Chromatogr. A* 2010, **1217**(26):4300–4305.

86. Vincent D, Wheatley MD, Cramer GR. **Optimization of protein extraction and solubilization for mature grape berry clusters.** *Electrophoresis* 2006, **27**(9):1853–1865.

87. Di Bernardo G, Del Gaudio S, Galderisi U, Cascino A, Cipollaro M. **Comparative evaluation of different DNA extraction procedures from food samples.** *Biotechnol Prog.* 2007, **23**(2):297–301.

88. Di Bernardo, G, Galderisi U, Cipollaro M, Cascino A. **Methods to improve the yield and quality of DNA from dried and processed figs.** Biotechnol Prog. 2005, **21**(2):546–9.

89. Kirsh S, Fourdrilis S, Dobson R, Scippo ML, Maghuin-Rogister G, De Pauw E: **Quantitative methods for food allergens: a review.** *Anal. Bioanal. Chem.* 2009, **395**(1):57–67.

90. Scaravelli E, Brohée M, Marchelli R, van Hengel AJ. **The effect of heat treatment on the detection of peanut allergens as determined by ELISA and real-time PCR.** *Anal Bioanal Chem.* 2009, **395**(1):127–37.

91. Iniestoa E, Jiméneza A, Prietoa N, Cabanillasc B, Burbanob C, Pedrosab MM, Rodríguez J, Muzquizb M, Crespoc JF, Cuadradob C, Linaceroa R. **Real time PCR to detect hazelnut allergen coding sequences in processed foods.** Food Chemistry 2013, **138** (2–3):1976–1981.

92. King CE, Debruyne R, Kuch M, Schwarz C, Poinar HN. **A quantitative approach to detect and overcome PCR inhibition in ancient DNA extracts.** *BioTechniques* 2009, **47**(5):941–949.

93. Henry A. Erlich. PCR technology. **Principles and applications for DNA amplification.** Macmillan Publishers; New York, Stockton press 1989.

94. Marmiroli N, Peano C, Maestri E. **Advanced PCR techniques in identifying food components.** In Food Authenticity and Traceability (pp 3–33). Cambridge, United Kingdom: Woodhead Publishing, 2003.

95. Holzhauser T, Wangorsch A, Vieths S. **Polymerase chain reaction (PCR) for detection of potentially allergenic hazelnut residues in complex food matrices.** *Eur Food Res Technol* 2000, **211**(5):360–365.

96. Rossi S, Scaravelli E, Germini A, Corradini R, Fogher C, Marchelli R. **A PNA-array platform for the detection of hidden allergens in foodstuffs.** *Eur Food Res Technol* 2006, **223**(1):1–6.

97. Piknová L, Pangallo D, Kuchta T. **A novel real-time polymerase chain reaction (PCR) method for the detection of hazelnuts in food.** *Eur Food Res Technol* 2008, **226**(5):1155–1158.

98. Brezna B, Hudecova L, Kuchta T. **A novel real-time polymerase chain reaction (PCR) method for the detection of walnuts in food.** *Eur Food Res Technol* 2006, **223**(3):373–377.

99. Mustorp S, Engdahl-Axelsson C, Svensson U, Holck A. **Detection of celery (*Apium graveolens*), mustard (*Sinapsis alba, Brassica juncea, Brassica nigra*) and sesame (*Sesamum indicum*) in food by real-time PCR.** *Eur Food Res Technol* 2008, **226**(4):771–778.

100. Turci M, Sardaro Maria Luisa S, Visioli G, Maestri E, Marmiroli M. **Evaluation of DNA extraction procedures for traceability of various tomatho products.** *Food Control* 2010, **21**:142–149.

101. Di Bernardo G, Del Gaudio S, Galderisi U, Cascino A, Cipollaro M. **Comparative evalu-ation of different DNA extraction procedures from food samples**. *Biotechnology Prog-ress*, 2007, 23: 297–301.
102. Peano C, Samson MC, Palmieri L, Gulli M, Marmiroli N. **Qualitative and quantitative evaluation of the genomic DNA extracted from GMO and non- GMO foodstuff with four different extraction methods**. *Journal of Agriculture and Food Chemistry*, 2004, **52**(23): 6962–6968.
103. Di Bernardo G, Galderisi U, Cipollaro M, Cascino A. **Methods to improve the yield and quality of DNA from dried and processed figs**. *Biotechnology Progress*, 2005, **21**(2): 546–549.
104. Venkatachalam M, Sathe S. **Chemical composition of selected edible nut seeds**. *J. Agric. Food Chem*, 2006, **54**(13):4705–4717.
105. Matthias B. **Determination of allergens in foods**. *Trends in Analytical Chemistry* 2001, **20**(11):662–672.
106. Poms RE, Anklam E, Kuhn H. **Polymerase chain reaction techniques for food allergen detection**. *Journal of AOAC International* 2004, **87**(6):1391–1397.

Chapter 5
Food Allergens Digestibility

Contents

Abbreviations

Acp	Epsilon-aminocaproic acid
AGE	Advanced glycation end products
AG-ONSu	N-hydroxysuccinimide ester of the amylose-glycylglycine adduct
ALGO	Alginic acid oligosaccharide
APC	Antigen presenting cells
APC	Antigen-presenting cell
BAT	Basophil activation test
BLG	β-lactoglobulin
BSA	Bovine serum albumin
CD	Circular dichroism
CHS	Chitosan

T. Ćirković Veličković, M. Gavrović-Jankulović, *Food Allergens,*
Food Microbiology and Food Safety, DOI 10.1007/978-1-4939-0841-7_5,
© Springer Science+Business Media New York 2014

CMD   Carboxymethyl dextran
CN   Casein
DG-ONSu   N-hydroxysuccinimide ester of the dextran-glycylglycine adduct
DIECA   Diethyldithiocarbamic acid
EAST   Enzyme allergosorbent test
EGTA   Ethylene glycol tetraacetic acid
ELISA   Enzyme-linked immunosorbent assay
GIT   Gastrointestinal tract
GMO   Genetically modified organisms
IFN-γ   Interferon γ
KCNO   Potassium cyanate
LTP   Lipid transfer proteins
mDC   Myeloid dendritic cells
NMR   Nuclear magnetic resonance
nOVA   Nitrated ovalbumin
nsLTPs   Non-specific lipid transfer proteins
OVA   Ovalbumin
OVM   Ovomucoid
PBMC   Peripheral blood mononucleated cell
PF   Peanut flour
POD   Peroxidases
PPO   Polyphenol oxidases
PR protein   Pathogenesis-related protein
RA-2S albumin   Rreduced and alkylated 2S albumin
RAST   Radioallergosorbent test
RBL   Rat basophilic leukaemia
SGF   Simulated gastric fluid
SIF   Simulated intestinal fluid
SKTI   Leguminous Kunitz-type inhibitor
SR   Scavenger receptor
TG   Transglutaminase
VMA   Copolymer of N-vinyl pyrrolidone and maleic anhydride
WDEIA   Wheat-dependent, exercise-induced anaphylaxis
WPC   Whey protein concentrate

Summary Although the exact mechanisms by which food allergens sensitize an individual remain currently unclear, most of them are thought to sensitize via the gastrointestinal tract (GIT). Resistance to proteolysis in the GIT would allow the allergens to maintain their immunogenic and allergenic motifs, and thus to interact with the immune system associated with the gastrointestinal epithelia, thereby inducing both sensitization and systemic allergy symptoms. For this reason, digestibility and gut permeability of food antigens are important factors of food allergenic potential. Testing of food protein susceptibility to digestion in simulated conditions of the GIT is often an unavoidable part of a food allergen characterization. Although the correlation between digestion stability of individual food allergens in simulated gastric fluids and their allergenicity may not be absolute for various reasons, there is a consensus in scientific community that for a food protein to be an oral sensitizer, it must have features which preserve its structure from degradation in the GIT, thus allowing intact allergen, or its larger fragments, to survive digestion in an immunologically active form to be taken up in the gut and sensitize the immune system. In the second phase of allergic reaction, allergens able to survive digestion in the GIT are responsible for clinical manifestations of food allergy, which may also include systemic and life-threatening anaphylaxis. By contrast, food allergens cross-reactive to respiratory allergens are often labile to digestion, and clinical symptoms induced by those are mild and related to oral cavity.

Digestion susceptibility of food allergens that sensitize via the GIT is inherently related to protein structural features. However, physiological changes in the digestion process, pathological conditions affecting digestion, as well as procedures and food-processing conditions that affect protein structure, food matrix, and food microstructure may all have a profound effect on the sensitizing potential and allergenicity of food proteins.

5.1   Classification of Allergens Based on Digestion Stability: Complete and Incomplete Food Allergens

5.1.1   *Protein Digestion in the Gastrointestinal Tract*

The process of protein digestion in the gastrointestinal tract (GIT) is carried out by three principal digestive proteases: pepsin, chymotrypsin, and trypsin. Pepsin starts digestion of proteins in the stomach, in the acidic environment. Pancreas secretes chymotrypsin and trypsin, powerful proteases, that contribute to protein digestion in the small intestine.

Pepsin is an acidic endoprotease, secreted by the stomach and it is active in the presence of hydrochloric acid. Pepsin's primary site of synthesis and activity is in the stomach (pH 1.5–2). Pepsin exhibits maximal activity at pH 2.0 and is inactive at pH 6.5 and above. However, pepsin is not fully denatured or irreversibly inactivated until pH 8.0. The enzyme has a preference for hydrophobic amino acids on either side of the scissile bond. There is a specificity for leucine, tyrosine, tryptophane, isoleucine, and glutamate in the P_1 position (the carboxyl side of the scissile peptide bond), and for tryptophane, isoleucine, tyrosine and phenylalanine in the P_1' position.

Chymotrypsin and trypsin are serin-proteases and main components of pancreatic juice, secreted by pancreas via the pancreatic duct to the first part of the intestine (duodenum). Both enzymes have an optimal pH of about 7.5–8.5. Chymotrypsin preferentially cleaves peptide amide bonds where the carboxyl side of the amide bond (the P_1 position) is a large hydrophobic amino acid (tyrosine, tryptophan, and phenylalanine). These amino acids contain an aromatic ring in their side chain that fits into a hydrophobic substrate-binding pocket (the S_1 position) of the enzyme. Trypsin has a preference for basic amino acids in the P1 position (lysine and arginine).

It has long been observed that many food allergens exhibit proteolytic stability, especially to pepsin digestion. Astwood et al. first conducted a systematic study to compare allergenicity and pepsin stability in simulated gastric fluid (SGF) of various allergenic and nonallergenic proteins. The results of the study led the authors to conclude that there is a strong relationship between allergenicity of a food protein and its stability to pepsin digestion [1]. That study has been criticized in later years [2–4], as the relationship between the stability of proteins to pepsin digestion in SGF and allergenicity has been inconsistent among studies conducted with reference to allergenic and nonallergenic proteins. It should be noted that digestion stability of a protein also depends on the composition of the digestion fluid and pepsin concentration and activity, parameters that often varied among studies, thus unavoidably leading to inconsistency of these data in the literature.

5.1.2 Classification of Food Allergens Based on Stability to Digestion

Food allergens that have both the ability to induce the immune system to produce high-affinity antibodies, particularly of the immunoglobulin E (IgE) class (to sensitize), and to elicit an allergic reaction (i.e., to trigger allergic symptoms in a sensitized subject) have been denominated as true or complete food allergens. There is another group of food allergens termed incomplete or non-sensitizing elicitors capable of eliciting allergic reaction, but not capable of inducing allergic sensitization [5]. These allergens are easily degraded during the gastrointestinal digestion and, as a result, they cannot sensitize directly.

For a complete food allergen, ability to reach immune cells is reflected in its stability to digestion. Allergens have no characteristic structural features other than that they need to be able to reach immune cells [6, 7]. Certain IgE-binding epitopes on food allergens are conformational [6, 7] and thus the epitope profile of a globular protein may be influenced by the microenvironment. Changes in pH, ionic strength, or binding of allergens to other molecules may affect the number of epitopes that are accessible for the antibodies by tightening or loosening the protein structure. In the latter case, new epitopes may be observed. Such epitopes are believed to be very prominent after proteolytic cleavage of the protein that occurs during digestion.

Majority of plant food allergens sensitizing via the GIT are either protective or storage proteins. Those proteins that can trigger the development of an allergic response through the GIT belong mainly to prolamin and cupin protein superfamilies. Prolamin and cupin superfamilies of proteins are characterized by remarkable structural stability that is also reflected in their resistance to heat and digestion [8]. The structural compactness of the prolamin superfamily is attributed to the presence of a conserved skeleton of cysteine residues that form four disulfide bridges [9, 10]. The proteins of the cupin superfamily possess a common β-barrel structure that appears to be a remarkably stable structural motif, resistant to both thermal denaturation and proteolysis [11, 12] (Table 5.1).

Cross-reactive food allergens involved in latex-fruit/vegetable or pollen-fruit/ vegetable syndromes are typical examples of non-sensitizing elicitors of food reactions. Patients with these allergies generally suffer mild symptoms, mostly limited to the oral cavity, the so-called oral allergy syndrome. Typical examples are members of the Bet v 1 homolog family (pathogenesis-related protein-10 family), such as the apple Mal d 1, the hazelnut Cor a 1.04, and the celery Api g 1, which are very labile to pepsin digestion. Hazelnut seems to contain either complete or incomplete food allergens. Allergens Cor a 1.04 (Bet v 1 homolog family) and Cor a 2 (profilin family) have been described to cross-react with birch pollen allergens [18], whereas hazelnut allergens with the ability to cause food allergy with more severe symptoms without primary pollen sensitization have also been reported [19].

5.2 Protein Conformational Stability, Digestibility, and Immune Response

5.2.1 Digestion Stability of Common Plant Food Allergens

5.2.1.1 Panallergens: Lipid Transfer Proteins, Profilins, Bet v 1 Homologs

Pollen-related food allergy is the most common food hypersensitivity in the adult population. Birch pollen-related allergy to Rosaceae fruits and vegetables, such as hazelnut, celery, and carrot, affects many patients in central and northern Europe. The presence of homologous allergens in pollens and fruit and vegetables, mainly members of the Bet v 1 and profilin families, is responsible for such cross-sensitizations. IgE recognition of panallergens, having highly conserved sequence regions, structure, and function, and shared by inhalant and food allergen sources, is often observed. Both profilins and Bet v 1 homologs in fruits/vegetables are labile allergens, readily hydrolyzed by SGF, and thereby candidates for pollen-related food hypersensitivity [20, 21]. The population allergic to Rosaceae fruits in the Mediterranean area presents a different clinical and sensitization pattern. More than 20 % of patients show fruit allergy without related pollinosis. Nonspecific lipid transfer proteins (nsLTPs) have been identified as the major allergens in this population [22, 23].

Nonspecific lipid transfer proteins are stable and highly conserved proteins of around 9–10 kD. Different degrees of sequence identity (from 30–95 %) are found between members of the family from different species. However, eight conserved cysteines forming four disulfide bridges, which are responsible for the lipid transfer protein (LTP) compact folding, are present in all members. Their common structural features, basic isoelectric point and high similarity in amino acid sequence, are the basis of allergic clinical cross-reactivity. This has been demonstrated for the LTP

Table 5.1 Stability of some food allergens in simulated gastric fluid

Source	Allergen name	Protein family/function	SGF stability	Reference
Cow's milk	α-casein	–	0	[3]
	β-lactoglobulin	Lipocalin	120	[3]
	bovine serum albumin (BSA)	Serum albumin	0	[3]
Egg	α-lactalbumin	Lysozyme c	0	[3]
	Lactoperoxidase	Myeloperoxidase	0	[3]
	ovalbumin	Serpin	5	[3]
	Ovomucoid	Kazal proteinase inhibitor homology (trypsin inhibitor)	0	[3]
Shrimp	Conalbumin	Transferrin	0	[3]
	lysozyme	Lysozyme c	60	[3]
	tropomyosin	Tropomyosin (contractile proteins)	0	[3]
Soy	β-conglycinin Beta-conglycinin (α-subunit)	Cupin superfamily (7S globulins)	0	[3]
			120	[3]
	β-conglycinin (β -subunit) Gly m 1 Gly m 1	Cupin superfamily (7S globulins)	2	[3]
			120	[3]
	Soybean Kunitz trypsin *inhibitor* (SKTI)	Prolamin superfamily (non-specific lipid transfer protein)	5	[3]
	Soybean lectin	Leguminous Kunitz-type inhibitor (trypsin inhibitor)		
		Leguminous lectin		
Peanut	Ara h 1	Cupin superfamily (7S globulins)	5	[3]
	Ara h 2		0.5	[3]
	Peanut lectin	Conglutin (2S seed storage albumins) lectin	5	[3]
Brazil nut	Ber e 1	Prolamin superfamily (2S seed storage protein)	15[a]	[13]
Sesame seed	Ses i 1	Prolamin superfamily (2S seed storage protein)	120[b]	[14]
Sunflower seed	SFA-8	Prolamin superfamily (2S seed storage protein)	30[a]	[13]
Mustard	Sin a 1	Prolamin superfamily (2S seed storage protein)	60[c]	[1]
	Bra j 1		60[c]	[1]
		prolamin superfamily (2S seed storage protein)		
Potato tuber	Patatin	Patatin	0	[3]
Peach	Pru p 3	Prolamin superfamily (non-specific lipid transfer protein)	30[d]	[15]

Table 5.1 (continued)

Source	Allergen name	Protein family/function	SGF stability	Reference
Melon	Cuc m 2	Profilin	0[c]	[16]
Apple	Mal d 1	Bet v 1 homolog	0.5[c]	[17]

SGF simulated gastric fluid
Pepsin/allergen ratio (w/w) = 13, except in:
[a] Pepsin/allergen ratio (w/w) = 17
[b] Pepsin/allergen ratio (w/w) = 0.05 (w/w) and pH 2.5
[c] Pepsin/allergen ratio (w/w) = 19
[d] Pepsin/allergen ratio (w/w) = 20

allergens of the *Prunoideae* subfamily, whose similarity is about 95 %, as demonstrated for the purified allergens of peach, apricot, plum, and apple. Sequence homology of LTPs of botanically unrelated foods has also been reported, as demonstrated for LTPs of maize and peach [24].

The experimental evidence points to their role in plant defense mechanisms against pathogens. LTPs are now classified in the pathogenesis-related (PR) protein families as PR-14 [23]. New lines of evidence suggest that the rigidity of the LTP scaffold is responsible for their resistance to proteolysis.

The structure and stability of the allergenic nsLTP of peach were compared with the homologous LTP1 of barley. The proteins were resistant to gastric pepsinolysis and were only slowly digested at 1–2 potential tryptic and chymotryptic cleavage sites under duodenal conditions. Molecular dynamics simulations of the proteins under folded conditions showed that the backbone flexibility is limited. These proteins were also characterized by nuclear magnetic resonance (NMR) spectroscopy at pH 1.8. This showed that the helical regions of both proteins remained folded at this pH. NMR hydrogen exchange studies confirmed the rigidity of the structures at acidic pH, with barley LTP1 showing some regions with greater protection. Gastroduodenal digestion conditions do not disrupt the three-dimensional (3D) structure of peach LTP, explaining why LTPs retain their ability to bind IgE after digestion and hence their allergenic potential [25].

Due to its compact fold and extreme resistance to heat treatments and pepsin digestion, LTPs are potentially severe food allergens, and are candidates for oral route sensitization [26]. Stability to heat treatments also implies the presence of active allergen forms in processed foodstuffs and beverages.

A specific geographical distribution pattern of sensitization to LTP allergens has been revealed. This allergen family is particularly important in the Mediterranean area, but shows a very limited incidence in central and northern Europe [23]. Sensitization to LTP has been recognized as a risk factor inducing allergic reactions in hazelnut allergy (i.e., Cor a 8 allergen of hazelnut is an nsLTP) [27] and allergy to maize, mustard, lettuce, and asparagus [23, 28]. It has been shown that maize LTP maintained its IgE-binding capacity after heat treatment, thus being the most eligible candidate for a causative role in severe anaphylactic reactions to both raw and cooked maize [29].

Severe grape allergy has been linked to LTP sensitization, as the allergenic activity of grape LTP was highly resistant to in vitro digestion. This property might facilitate sensitization through the GIT and might also potentiate the ability of LTPs to elicit severe allergic reactions in sensitized individuals [30]. Apple LTP is also resistant to simulated enzymatic digestion [31]. Resistance to SGF and to acidic proteases of fungal origin has been shown for allergenic LTPs from carrot [31] and barley [32].

Peach allergen Pru p 3 has been extensively studied in simulated conditions of the GIT. The products of the digestion were essentially derived from trypsin action, whereas the protein appeared to be resistant to pepsin and chymotrypsin. The identified peptides could be classified as low molecular weight and high molecular weight peptides. The latter consisted of the full-length protein, with the disulfide bridges still intact, deprived of the smaller peptides [33].

LTP has been identified as an important allergen of cherry fruits [34]. In a study that compared stability of natural and recombinant cherry allergens Pru av 3 (nsLTP), Pru av 1 (Bet v 1 homolog), and Pru av 4 (profilin) to pepsin digestion and to thermal processing, it has been demonstrated that LTPs showed the highest resistance to digestion by pepsin (Pru av 3 > Pru av 1 > Pru av 4). Immunologically active Pru av 3 was detectable after 2 h of digestion by pepsin, whereas IgE reactivity of Pru av 1 and Pru av 4 was abolished within less than 60 min. In contrast with Pru av 1, IgE reactivity to nsLTPs was not diminished in thermally processed fruits, and secondary structures of purified Pru av 3 were more resistant to heating [34].

Bet v 1 homolog allergens from apples (Mal d 1), hazelnuts (Cor a 1.04), and celery (Api g 1) were tested for pepsin and trypsin digestion [35]. Pepsin completely destroyed IgE binding of all allergens within 1 s, and trypsin completely destroyed IgE binding of all allergens within 15 min, except for the major hazelnut allergen, which remained intact for 2 h of trypsinolysis. Allergens after gastrointestinal digestion did not induce basophil activation, but induced proliferation in peripheral blood mononucleated cells (PBMCs) from allergic and nonallergic individuals. Digested Mal d 1 and Cor a 1.04 still activated Bet v 1-specific T cells, whereas digested Api g 1 did not. Thus, gastrointestinal degradation of Bet v 1-related food allergens destroys their histamine-releasing, but not T-cell-activating, property [35].

Four recombinantly produced Bet v 1 homologous food allergens from peach (Pru p 1), celery (Api g 1), apple (Mal d 1), and hazelnut (Cor a 1) were used to probe the structural responsiveness of the Bet v 1 scaffold to gastric digestion conditions [17]. Low pH induced conformational changes of all Bet v 1 homologs. The homologs were rapidly digested by pepsin, losing their IgE-binding activity, although the kinetics and patterns of digestion varied subtly between homologs, Api g 1 being the most stable. Gastric phosphatidyl-choline induced conformational changes in all homologs, but only Mal d 1 penetrated the phosphatidyl-choline vesicles, slowing its digestion and retaining more of its allergenic activity [17]. Thus, the Bet v 1 scaffold is susceptible to acidic pH of the gut and pepsinolysis and interacts with phosphatidyl-choline vesicles, properties which can explain effects of the gastric environment on their allergenicity.

Bet v 1 homolog allergen in peanut (Ara h 8) is an allergen of minor importance, and similarly to other Bet v 1 homolog proteins, it is not stable to roasting and to gastric digestion [36].

Profilin is a highly conserved protein in pollen and vegetable food. Homologous proteins from different sources are highly cross-reactive. Profilin is a well-defined plant panallergen, showing prevalences around 30 % in fruits and vegetables. In patients with oral allergy syndrome, a 13-kDa protein, identified as a profilin (Cuc m 2), is a major melon allergen highly susceptible to pepsin digestion [16, 37]. Orange profilin Cit s 2, unlike other plant food profilins, is a major and highly prevalent allergen [38]. A peanut profilin, identified as an allergen Ara h 5, shows a 3D structure similar to birch pollen profilin, Bet v 2, [39]. Profilin is also an important allergen in pollen-related kiwifruit allergy [40]. Kiwifruit allergic patients from southern Europe were mainly sensitized to Act d 9 (profilin, 31 %) and Act d 10 (nsLTP, 22 %) [41].

Profilin was identified as a new mustard allergen (Sin a 4) [28]. The resistance to proteolysis and heating of the yellow mustard allergens Sin a 1 (2S albumin), Sin a 3 (nsLTP), and Sin a 4 (profilin) was examined in order to explain their potential capability to induce primary sensitization by ingestion. Sin a 1 and Sin a 3 resisted gastric digestion showing no reduction of the IgE reactivity. Intestinal digestion of Sin a 1 and Sin a 3 was limited, and proteins retained significant IgE-binding reactivity. Sin a 1 and Sin a 3 were stable to heating. These two allergens would be therefore able to sensitize by ingestion. Mustard profilin (Sin a 4) was completely digested by gastric treatment, and its conformational structure was modified at 85 °C [42].

5.2.1.2 2S Albumin Storage Proteins

2S albumin storage proteins have been reported as major food allergens in seeds of many mono- and dicotyledonous plants. 2S albumins are considered to sensitize directly via the GIT. The high stability of their rigid intrinsic protein structure, dominated by a well-conserved skeleton of cysteine residues, to the harsh conditions present in the GIT suggests that these proteins are able to cross the gut mucosal barrier to sensitize the mucosal immune system and/or elicit an allergic response. Finally, the interaction of these proteins with other components of the food matrix might influence the absorption rates of immunologically reactive 2S albumins, but also the immune response they elicit [43].

Among the peanut-allergenic proteins, Ara h 2 is one of the most commonly recognized allergen. Ara h 2 is a 17-kDa protein that has eight cysteine residues that could form up to four disulfide bonds. Upon treatment with trypsin, chymotrypsin, or pepsin, a number of relatively large fragments are produced that are resistant to further enzymatic digestion. These resistant Ara h 2 peptide fragments contain intact IgE-binding epitopes and several potential enzyme cut sites that are protected from the enzymes by the compact structure of the protein [44].

The allergenic 2S albumin Ara h 2 and the homologous allergen Ara h 6 were studied at the molecular level with regard to allergenic potency of native and protease-treated allergens. Both allergens contain cores that are highly resistant to proteolytic digestion and high temperature [45]. Even though IgE antibody-binding capacity was reduced by protease treatment, the mediator release from a functional equivalent of a mast cell or basophile, the humanized rat basophilic leukemia (RBL) cell, demonstrated that this reduction in IgE antibody-binding capacity does not necessarily translate into reduced allergenic potency. Native Ara h 2 and Ara h 6 have virtually identical allergenic potency as compared with the allergens that were treated with digestive enzymes. The folds of the allergenic cores are identical with each other and with the fold of the corresponding regions in the undigested proteins. The extreme immunological stability of the core structures of Ara h 2 and Ara h 6 provides an explanation for the persistence of the allergenic potency even after food processing [46]. The major 2S albumin allergen from Brazil nuts, Ber e 1, was subjected to gastrointestinal digestion using a physiologically relevant in vitro model system either before or after heating (100 °C for 20 min). The characteristic conserved skeleton of cysteine residues of 2S albumin family and, particularly, the intrachain disulfide bond pattern of the large subunit play a critical role in holding the core protein structure of Ber e 1 together even after extensive proteolysis, and the resulting structures still contain potentially active B- and T-cell epitopes [47].

Ses i 1 from white sesame seeds has been shown to be extremely stable to acid conditions, thermal processing, and in vitro gastrointestinal digestion. Although Ses i 1 unfolded to a limited extent on heating, it refolded on cooling to an almost native structure due to disulfide bonds that increase the stability of the folded protein. This structural stability could explain the fact that Ses i 1 digestion was not affected by preheating at 100 °C, either at acid or neutral pH. Following in vitro gastric digestion, Ses i 1 remained completely intact. Hence, Ses i 1 may be able to retain both linear and conformational epitopes following severe heat treatments in order to sensitize an individual or provoke an allergic reaction in a sensitized individual [14].

5.2.1.3 Prolamins

Ara h 1, a major peanut allergen, is not very resistant to gastric digestion. In vitro digestion of Ara h 1 with pepsin and porcine gastric fluid resulted in virtually identical hydrolysis patterns. Protein digestion in the porcine stomach is carried out exclusively by pepsin [48]. Ara h 1 contains 23 IgE-binding sites that are evenly distributed along the linear sequence of the molecule [49]. It has been shown that after digestion by the gastrointestinal enzymes, large proteolytic fragments of Ara h 1 still contain multiple IgE-binding epitopes and retain their allergenic potential [50]. Ara h 1 forms homotrimers that could provide certain protection from protease digestion and denaturation [51]. Individual patient-specific epitope patterns have been identified for the major allergen Ara h 1. IgE-binding epitopes have been suggested as biomarkers for persistency and severity of allergy to peanut, wherefore recognition of particular epitope patterns, or motifs, could be a valuable tool for prevention, diagnosis, and treatment of food allergy [52].

5.2.1.4 Thaumatin-Like Proteins

The family of thaumatin-like proteins (TLPs) belongs to PR-5 family of proteins and it is described as an important allergen of fruits [53]. Many of the described TLPs are resistant to gastric digestion [54–56]. By contrast, kiwi TLP is not resistant to gastric digestion [57]. In a study conducted to compare digestibility of TLPs of kiwi, banana, apple, and sour cherry, the only gastric labile protein was kiwi TLP [58]. However, kiwi TLP is an important allergen of kiwifruit [57].

5.2.2 Stability of Animal Allergens to Digestion

The most important animal food allergens are present in milk, egg, and seafood. Animal food allergens can be classified into three main families—tropomyosins, EF hand proteins, and caseins (CNs)—as well as a long tail of families containing only one to three reported allergens in each. For all three main animal allergen families, their ability to act as allergens seems to be related to their closeness to human homologs. Proteins with a sequence identity of up to 54% to human homologs were all allergenic, whereas those with a sequence identity greater than 63% to human homologs were rarely allergenic. This observation probably relates to the requirement of proteins to be recognized as nonself to mount an immune response, and it has been argued that a low degree of similarity to a host's proteome is required for immunogenicity.

5.2.2.1 Egg Allergens' Stability to Digestion

Hen egg white is comprised of a complex mixture of proteins, which greatly differ in their physicochemical characteristics and relative abundance. Egg allergens have been described in both white and yolk, and the egg white proteins, ovomucoid (OVM), ovalbumin (OVA), ovotransferrin, and lysozyme, have been adopted in the allergen nomenclature as Gal d 1- d 4 [59]. Major allergens of egg white are all easily digested by pepsin (Table 5.1). However, experimental data show that peptides generated by digestion may still remain immunologically active [60]. Egg white exhibited residual immunoreactivity after gastrointestinal digestion due to the presence of intact OVA and lysozyme, as well as due to several IgE-binding peptides derived from OVA. It has also been shown that the presence of egg yolk slightly increased the susceptibility to hydrolysis of egg white proteins. However, the resultant immunoreactivity against IgE of egg white proteins after in vitro digestion was not significantly modified by the presence of yolk components.

OVM was digested in SGF in order to examine the reactivity of the resulting fragments to IgE in sera from allergic patients. OVM was first cleaved near the end of the first domain, and the resulting fragments were then further digested into smaller fragments. When the digestion of OVM was kinetically analyzed, 21 % of the examined patients' sera retained their IgE-binding capacity to the small 4.5-kDa fragment [61].

5.2.2.2   Cow's Milk Allergens' Stability to Digestion

Important cow's milk allergens are CNs (Bos d 8), β-lactoglobulin (Bos d 5), α-lactalbumin (Bos d 4), and bovine serum albumin (BSA; Bos d 6) [62, 63]. CN are very labile in the gastric fluid, as well as Bos d 4 and Bos d 6 (Table 5.1). A major milk allergen Bos d 5 has a compact globular structure, whose conformation and even aggregation states depend highly on the pH and temperature of solution [64]. A shift in secondary structures can also be observed for this compact protein at different pH values. Stabilized by several disulfide bonds, the structure of Bos d 5 is so compact that it is known as one of the most resistant proteins to pepsin digestion (Table 5.1) [3].

The kinetics of breakdown of the bovine milk allergen alpha-lactalbumin during in vitro gastrointestinal digestion was found to be altered by interactions with physiologically relevant levels of phosphatidylcholine, a surfactant that is abundant in milk and is actively secreted by the stomach. Breakdown during gastric digestion was slowed in the presence of phosphatidylcholine and accompanied by small alterations in the profile of resulting peptides [65].

A biochemical model of infant gastroduodenal digestion, having reduced levels of protease, phosphatidylcholine, and bile salts compared with the adult model, has been developed in order to study comparative resistance of β-casein and β-lactoglobulin in infant and adult digestion models. β-Casein was digested more slowly using the infant model compared with the adult conditions. β-Lactoglobulin was more extensively degraded in the infant model compared with the adult one. This difference was attributed to the tenfold reduction in phosphatidylcholine concentration in the infant model, limiting the protective effect of this phospholipid on β-lactoglobulin digestion [66].

5.2.2.3   Fish and Seafood Allergens' Stability to Digestion

In fish, the dominating allergen is the homolog of Gad c 1 from cod, formerly described as protein M. A close cross-reactivity exists within different species of fish between members of this EF hand calcium-binding protein family, denominated the parvalbumins. This cross-reactivity has been indicated to be of clinical relevance to several species, since patients with a positive double-blind, placebo-controlled food challenge to cod will also react with other fish species, such as herring, plaice, and mackerel [59].

Parvalbumins are also the major allergens in commonly consumed tropical fish. A study evaluated the allergenicity of four commonly consumed tropical fish, the threadfin (*Polynemus indicus*), Indian anchovy (*Stolephorus indicus*), pomfret (*Pampus chinensis*), and tengirri (*Scomberomorus guttatus*) [67]. The major allergen of the four tropical fish was the 12-kDa parvalbumin. IgE cross-reactivity of these allergens to Gad c 1 was observed to be moderate to high in the tropical fish studied. They were cross-reactive with each other, as well as with Gad c 1. Using the monoclonal anti-parvalbumin antibody, the presence of monomeric and oligomeric parvalbumin was demonstrated in several fish analyzed [68].

Lep w 1, a major allergen of whiff fish, is a calcium-binding beta-parvalbumin. Purified Lep w 1 was thermally stable up to 65 °C, but only when calcium was bound as a ligand and the tests were performed at neutral pH. In contrast to Gad m 1, Lep w 1 lost its structure completely when calcium was depleted from the protein. A partial loss of structure was also observed at acidic pH; however, the allergen retained its full IgE-binding ability. The partially denatured Lep w 1 was easily digested by pepsin within 2 min. The circular dichroism (CD) measurements showed a partial denaturation of the protein at pH 2.5, which could explain the efficient pepsinolyis of Lep w 1 within seconds in in vitro gastric digestion assays. No difference of gastric stability was observed between the ethylene glycol tetraacetic acid (EGTA)-treated and untreated parvalbumin, as the chelator was inactive in the acidic SGF, pH 2.5. In contrast to a raw fish extract, the cooked extract showed higher resistance to pepsinolysis. Cooked fish extract fragments were still recognized by patients' IgE after more than 120 min of digestion by pepsin. Also, the antibody failed to recognize high-molecular mass protein bands in cooked fish, but detected a protein at 24 kDa. Such a molecular weight is characteristic of a parvalbumin dimer. The monomeric Lep w 1 was digested by pepsin after 5 s, whereas the dimer was still detected after 120 min of digestion. Despite the remarkable stability to heating, Lep w 1 was easily digested using physiological gastric conditions. However, food processing such as cooking could generate dimers that were partially stable towards gastric digestion. It is likely that the observation of stable parvalbumin dimers and the formation of protein aggregates after cooking explain the high allergenicity of this fish [69].

The major allergen of crab is tropomyosin, a myofibrillar protein that is composed of two identical subunits with molecular masses of 35–38 kDa and an isoelectric point of 4.5. In contrast with other myofibrillar proteins, tropomyosin is stable, it can tolerate heat and grinding, and resists harsh conditions of pepsin digestion for up to 15 min [70]. The digestibility of the purified mud crab allergen tropomyosin was examined by electrophoresis. Tropomyosin was relatively stable in SGF, as the digestion of tropomyosin by pepsin was gradual, and more proteolytic fragments appeared with increased digestion time, while the largest fragment with size of approximately 34 kDa was resistant to digestion [71].

Tropomyosins have also been described as major allergens of shrimps [72, 73]. Tropomyosins of different shrimps have been investigated and were found to be relatively stable to pepsin digestion [71, 74, 75]. The digestive stability of tropomyosin and other food proteins from mud crab were tested in SGF and simulated intestinal fluid (SIF) digestion assays. In SGF, proteins such as actin and the original band of myosin heavy chain were rapidly degraded within a short period of time, while tropomyosin was relatively resistant to pepsin digestion. In SIF, myosin heavy chain was easily decomposed, while tropomyosin and actin were similarly resistant to digestion. Further study by IgE immunoblotting and inhibition enzyme-linked immunosorbent assay (ELISA), using sera from crab-allergic patients, indicated that allergenicity of tropomyosin was partially decreased by digestion. Similarly, major allergen tropomyosin of Chinese mitten crab, Pacific white shrimp, and Grass prawn was relatively resistant to pepsin, while susceptible to trypsin and chymotrypsin digestion [71, 74, 75].

5.2.3 Correlation Between Results from In Vitro Digestion Tests and Allergenicity

Gastric Fluid Digestibility Assays Results of Fu and coworkers [3] suggest that food allergens with high allergenicity are not necessarily more resistant to SGF digestion than proteins with lower allergenicity. For example, members of the plant lectin and contractile protein groups showed similar SGF digestibility, irrespective of their allergenicity. The digestibility of shrimp tropomyosin, a major allergen, was similar to those of nonallergenic chicken, bovine, and pork tropomyosins. The shrimp tropomyosin formed four fragments that remained clearly visible up to 5 min of digestion. On the other hand, the pork, chicken, and bovine tropomyosins each formed a single fragment, which continued to degrade during the SGF reaction. It is not clear whether this difference in degradation pattern is of relevance with respect to the difference in allergenicity exhibited by these proteins. The digestive stability within the storage protein and enzyme categories, on the other hand, varied greatly. Within the storage protein group, there was no clear indication that food allergens were more resistant to SGF digestion than proteins with unproven allergenicity. A food allergen may be more stable, equally stable, or less stable to SGF digestion than proteins of unproven allergenicity [3].

There was no apparent trend to indicate that a protein with a higher percent allergenicity is more resistant to SGF digestion. α-Casein, a major milk allergen to which up to 100 % of patients have IgE, was degraded more rapidly in SGF than BSA, a minor milk allergen. The major egg allergen, OVA, with an allergenicity of 100 %, showed a lower SGF stability than the minor egg allergen, lysozyme. The major soybean allergen, Gly m 1, was found to be less stable to SGF digestion than the soybean trypsin inhibitor, a minor soybean allergen [3]. Similarly, Ara h 1 and Ara h 2, both major peanut allergens, showed a lower stability than soybean trypsin inhibitor. Shrimp tropomyosin and patatin, to which 82 % of shrimp-allergic individuals and 74 % of potato-allergic individuals, respectively, have IgE, also degraded rapidly.

Intestinal Fluid Digestibility Assays Food allergens were not necessarily more resistant to SIF digestion than proteins with unproven allergenicity. Some major food allergens showed rather rapid degradation in SIF. For example, *Ara h 2* degraded within 0.5 min; shrimp tropomyosin degraded instantly, although it formed peptide fragments that were stable for 0.5 min. The major milk allergen, β-lactoglobulin B, which was highly resistant to SGF digestion, was relatively labile to SIF digestion.

Some allergens (e.g., OVA, conalbumin, papain, and bromelain) that were labile to SGF digestion seem to be stable in SIF. However, this characteristic was not unique to allergens.

A number of nonallergenic enzymes that were readily digested in SGF (e.g., rubisco) also showed high stability in SIF. Some major allergens that were labile to digestion in SGF (e.g., shrimp tropomyosin) were also labile in SIF. Similar SIF stability seems to exist among members of certain protein groups, irrespective of their allergenicity. The plant lectins as a group showed a high stability to SIF, whereas the tropomyosins were relatively labile. The SIF stability within the storage protein group varied greatly, although some similarity could be observed among members of certain protein families. All the trypsin inhibitors tested showed similar SIF stability irrespective of their allergenicity. Both members of the papain superfamily, papain and bromelain, also showed similar SIF stability. A comparison of the SIF stability between major and minor allergens did not indicate a clear correlation between in vitro digestibility and protein allergenicity. Food allergens of high allergenicity were not necessarily more resistant to SIF digestion than allergens of low allergenicity. A number of major food allergens (e.g., α-casein, β-lactoglobulin, Ara h 2) showed less stability in SIF than some minor allergens (e.g., soybean lectin, peanut lectin, and lysozyme) [3].

There are also pepsin-sensitive proteins within the food allergens that are suspected to sensitize through the GIT, as is the case of shrimp tropomyosin or milk CNs, α-lactalbumin, and BSA. A possible explanation can be that the large stable proteolytic fragments generated during digestion have the potential to bind IgE and play a role in sensitization. Van Beresteijn et al. [76] found that the minimum molecular mass of whey peptides necessary to elicit an immunological response was between 3 and 5 kDa. Likewise, Huby et al. [77] proposed that an allergen must contain at least two IgE-binding sites or epitopes, each with a minimum length of 15 amino acid residues, in order to make possible the antibody binding.

On the other hand, the relative abundance of the allergen in the food should be another factor to be considered together with its structural stability, since abundance may influence the dose of allergen that survives gastrointestinal digestion. The major allergens found in foods, which include a large proportion of the human diet, such as milk, egg, fish, or potatoes, are all highly abundant, comprising 20–60 % of the protein in the original food.

Additionally, some proteins that are very stable to pepsin digestion have not been reported as allergens, such as zein from corn or concanavalin [3, 4, 78]. This stresses that, in addition to digestive stability, proteins should have the ability to stimulate the immune system in order to sensitize individuals and/or elicit an allergic reaction. Therefore, although the assessment of the resistance to the gastrointestinal digestion of proteins with the ability to sensitize individuals could provide valuable information on their potential allergenicity, no single criterion can be used to predict human allergenic potential of food proteins. The digestion resistance data of proteins should be interpreted in conjunction with other factors, such as specific serum screening or the sequence homology to known allergens [79].

5.3 Impact of Food Processing on Digestibility and Allergenicity of Food Allergens

5.3.1 Food Processing and Food Allergenicity

Processing procedures, food protein modifications, and food structure may modulate the allergenic properties of foods. Many of the food allergens are stable proteins that are very resistant to digestion by gastrointestinal enzymes [1, 2], or when digested, give peptide fragments of significant size which retain their IgE-binding and T-cell-stimulating capacities [50]. A special feature of food matrix is that it may contain substances inherently able to hamper digestion by enzymes, by making physical obstacles to enzyme action. Those described so far are lipids, i.e., phosphatidylcholine, which is secreted by the stomach and also abundant in milk [65], and polysaccharides, i.e., pectin, of fruit matrices [58, 80]. On the other hand, food matrix may also help the allergenic nature of food proteins by contributing to the activation of immune cells.

It was shown that purified peanut allergens, unlike a whole peanut extract, possess little intrinsic immune-stimulating capacity, and that the immune response to these allergens can be adjuvated by the presence of a food matrix. Soluble peanut allergens do not possess intrinsic adjuvanticity, and they therefore need an accompanying adjuvant (in the case of PE provided by the food matrix) to be able to activate antigen-presenting cell (APC) and to induce subsequent immune stimulation [81].

Endogenous Brazil nut lipids are required for the induction of optimal antibody responses to Ber e 1 in mice. Appropriate antibody-binding sites are present on both natural and recombinant forms of Ber e 1, suggesting that the impact of lipid is at the induction phase, rather than antibody recognition, and is possibly required for efficient antigen presentation [82].

When Gly m Bd 30 K was coadministered with dietary fats, absorption was enhanced. It was demonstrated that dietary fat or an exogenous emulsifier increased the gastrointestinal absorption of a major soybean allergen (Gly m Bd 30 K) in mice. Oil-body-associated protein has increased protease resistance in the stomach and also the transit time in the small intestine. Enhanced absorption of Gly m Bd 30 K from an epithelial cell membrane of the small intestine to the blood circulation allows sensitization [83].

Thereby, processing procedures also targeting the food matrix, such as action of lipases or glycohydrolitic enzymes, may potentially influence the digestibility and allergenicity of food allergens.

Application of high-pressure technology during or before the enzymatic processing may reduce the size of intermediate peptide fragments and improve the hypoallergenic properties of the treated proteins [84, 85]. By enzymatic action, peptides able to inhibit uptake of allergens may be produced [86].

5.3.2   Methods for Assessing Allergenicity of Modified Food Allergens

The efficacy of chemical and enzymatic methods used to alter food allergen reactivity must be verified by analyzing the treated allergen's (or treated allergenic food) ability to trigger an immune response. As the reactivity of an allergen is often described by its ability to bind IgE antibodies, reduced IgE activity may indicate a modification or removal of food allergen(s). A variety of rapid assay methods are used as analytical tools in research of molecular structure, integrity, and biological activity of food allergens and their epitopes. Allergen reactivity can be determined by in vitro, *ex vivo,* and in vivo testing. In vitro tests are often inexpensive, quick, and without a threat to human or animal subjects. On the other hand, in vivo assays provide a more accurate representation of the research. *Ex vivo* tests are advantageous, because they are measuring allergenic response on an effector cell level, using human subject's blood without exposing them to risk. Animal models may also be used, though these models are not always analogous to the human. However, animal models may be a useful tool for predicting sensitizing potential of proteins introduced into diet by genetic manipulation or modified allergens which may carry higher allergenic risk.

5.3.3   Modifications of Food Allergens

Modifications of food allergens may occur during food processing and preparation (i.e., sonication, high-pressure treatment, boiling, roasting, cooking, and baking), or as an attempt to create hypoallergenic food products. The former may engage other food components leading to unintended conjugations of food allergens with other food compounds, especially carbohydrates, while the latter often employs purified food allergens, or protein extracts, and investigates immunological properties of these preparations with a vision to improve food allergy treatment.

Most foods are subjected to thermal processing. Thermal processing provides many beneficial effects, but also may bring substantial changes in allergenicity. Thermal processing is as likely to increase allergenicity as to reduce it, especially via promotion of chemical modifications of allergens. These changes are highly complex and not easily predictable, but there are a number of major chemical pathways that lead to distinct patterns of modification. Perhaps the most important of these is through the reaction of protein amino groups with sugars, leading to a mixture of advanced glycation end product-modified protein derivatives (Maillard reaction products). Thus, heat-induced changes in allergens are often jointly investigated with the effects of the Maillard reaction modification of proteins.

5.3.3.1 Chemical Modifications of Proteins Used for Improvement of Nutritional Purposes

The aim of chemical modifications of proteins used for nutritional purposes is improvement of techno-functional properties of food proteins (solubility, emulsification, foaming, gelling, etc.) and preservation of their nutritional value. Chemical modifications of food proteins can lead to a change in the charge and hydrophobicity of proteins, which in turn can diminish or eliminate allergenicity of food allergens. However, regardless of the application of nontoxic reagents, chemical modifications are not so often applied in the food industry, as the procedures of removing remaining chemical agents may be complex and expensive.

Acylation Acylation of allergens by treatment with acid anhydrides, such as acetic or succinic acid anhydrides, blocks positively charged amino groups on the protein molecule, and the remaining free carboxyl groups of aspartic and glutamic acid residues make the net charge of the modified protein more negative. Chemical modifications of respiratory allergens, leading to preparations of reduced allergenicity, have been applied to various allergenic sources and demonstrated potential for creation of hypoallergenic allergen products for immunotherapy [87–89].

Similarly, Szymkiewicz and Jędrychowski [90] modified pea proteins with acetic or succinic anhydride. Immunoreactivity of albumins and legumin, as estimated by ELISA with rabbit polyclonal antibodies, reduced by 91–99% and 78–97% after succinylation and acetylation, respectively, while immunoreactivity of vicilin fraction reduced down to 12% and 17%, respectively.

Carbamoylation Mistrello et al. [91] chemically modified OVA by reaction with potassium cyanate (KCNO), which transforms the ε-amino group of the lysine of proteinaceous allergens into the ureido groups. KCNO-modified (carbamylated) allergens have low allergenic potency, as demonstrated in vitro and in vivo. When used to immunize rabbits, carbamylated allergens still induce IgG antibodies able to cross-react with native allergens.

Nitration Nitration of proteins on tyrosine residues, which can occur due to polluted air under "summer smog" conditions, has been shown to increase the allergic potential of respiratory allergens [92]. This posttranslational modification of proteins is likely to trigger immune reactions and provides a molecular rationale for the promotion of allergies by traffic-related air pollution. Since nitration of tyrosine residues is also observed during inflammatory responses, this modification can influence protein immunogenicity and might therefore contribute to food allergy induction. In a study by Untersmayr et al. [93], BALB/c mice were immunized intragastrically by feeding untreated OVA and nitrated ovalbumin (nOVA) with or without concomitant acid suppression. While oral immunizations of nOVA under antiacid treatment did not result in IgG and IgE formation, intraperitoneal immunization induced high levels of OVA-specific IgE, which were significantly increased in the group that received nOVA by injection. Furthermore, nOVA triggered significantly enhanced mediator release of effector cells of sensitized allergic animals. In gastric digestion experiments, nOVA degraded within few minutes, whereas OVA remained stable up to 120 min. Additionally, one tyrosine residue being very efficiently nitrated is part of an OVA epitope recognized exclusively after oral sensitization. These data indicated that despite the enhanced triggering capacity in existing allergy, nitration of OVA may be associated with a reduced de novo sensitizing capability via the oral route due to enhanced protein digestibility and/or changes in antibody epitopes. Although the authors considered effects of endogenously nitrated allergen, these results imply that nitration of food allergen as method has no potential for reducing its allergenic potential.

Polymerization by Glutaraldehyde By treatment with glutaraldehyde, aldehyde groups of glutaraldehyde react with the amino groups of protein resulting in cross-linked allergen proteins with altered immunological characteristics [94]. The ability of glutaraldehyde-treated allergens to stimulate T cells can also decrease following modification [95]. Yang et al. demonstrated that, whereas in vivo administration of OVA induces cytokine synthesis, administration of glutaraldehyde polymerized, high relative molecular weight OVA leads to a 20-fold increase in the ratio of interferon γ(IFN-γ)/interleukin-4 (IL-4) and IFN-γ/IL-10 synthesis observed after short-term, antigen-mediated restimulation directly ex vivo [96]. Thus, this approach allows selective activation of strongly type 1 T helper cell (Th1)-dominated immune responses to protein antigens, and it may be useful in clinical settings where the ability to actively select specific patterns of cytokine gene expression would be advantageous.

5.3.3.2   Modifications of Allergens via Maillard Reaction

The Maillard browning reaction occurs during heat processing when lysine residues in dietary protein chemically react with sugars.

The effect of the glycosylation, as a result of thermal treatment in the presence of glucose, on the digestibility and IgE binding of codfish parvalbumin was recently investigated [97]. The glycosylation of codfish parvalbumin did not affect the pepsin digestibility of parvalbumin, and the peptides resulting from this digestion showed low IgE binding, regardless of glycosylation. However, glycosylation of parvalbumin led to the formation of higher-order structures that were more potent IgE binders than native, monomeric parvalbumin. Therefore, authors concluded that food-processing conditions applied to fish allergen can potentially lead to increased allergenicity, even while the protein's digestibility is not affected by such processing [97].

Glycation of hazelnut 7S globulin Cor a 11 at 37 °C did not influence the specific IgE binding, which was influenced by heating at 60 and 145 °C. Heating at 145 °C, with or without glucose, increased basophil degranulation capacity of Cor a 11. This is possibly related to aggregation of the allergen as a result of the heat-promoted Maillard reaction [98].

Another study showed that heat treatment significantly reduced IgE binding to both OVA and OVM, whereas the Maillard reaction reduced the IgE binding to OVA, but increased IgE binding to OVM. In contrast, heat treatment significantly favored OVA digestibility, but glycation impaired it, and these treatments did not affect the digestibility of OVM. Thus, heat treatment and glycation by the Maillard reaction showed an influence on the allergenicity of the main egg white proteins, that could be related to their resistance to denaturation and digestive enzymes [99].

The Maillard reaction of squid tropomyosin and ribose decreased specific IgE binding to glycosylated tropomyosin. Pepsin digestion diminished the specific IgE-binding ability of both tropomyosin and glycosylated tropomyosin, and the reduction of the allergenicity by the Maillard reaction of tropomyosin with ribose remained after peptic digestion [100].

Scallop tropomyosin, the major allergen of shellfish, was prepared from adductor muscles and reacted with four reducing sugars to investigate the effect of the Maillard reaction on the allergenicity of the Maillard reaction. The IgE-binding ability of tropomyosin increased significantly with the progress of the reaction with glucose, ribose, and maltose, but not with maltotriose. The allergenicity was enhanced at the early stage of the Maillard reaction, and the trend of the effect depended on the type of reducing sugar used [101].

The Maillard reaction may also employ autologous polysaccharides. The effects of autologous plant polysaccharides on the immunoreactivity of buckwheat Fag t 3 (11S globulin), following the Maillard reaction, showed that the IgE-binding properties of Fag t 3 decreased dramatically, with significant changes also being observed in the electrophoretic mobility, secondary structure, and solubility of the glycated Fag t 3 [102]. Glycation that occurs via the Maillard reaction during the processing of buckwheat food may be an efficient method to reduce Fag t 3 allergenicity.

Promotion of the Maillard reaction also occurs during roasting. It has been shown that roasted peanuts have a higher level of IgE binding than raw peanuts [103]. This increase in IgE binding of roasted peanuts could be due to increased levels of protein-bound end products or adducts such as advanced glycation end products (AGEs), malondialdehyde, N-(carboxymethyl)lysine, and 4-hydroxynonenal. IgG antibodies to each of these adducts were produced in order to examine the levels of modifications in both raw and roasted peanuts. Results showed that adducts were all present in raw and roasted peanuts. Roasted peanuts exhibited a higher level of AGEs and malondialdehyde adducts than raw peanuts. IgE was partially inhibited in a competitive ELISA by antibodies to AGEs, but not by antibodies to other adducts, indicating that IgE has an affinity for peanut AGE adducts. Roasted peanuts exhibited a higher level of IgE binding, which was correlated with a higher level of advanced glycation end product adducts [104].

The glycation structures of AGEs are suggested to function as PR immune epitopes in food allergy, contributing to the enhanced immune response to glycated allergens. T-cell immunogenicity of food AGEs was identified by using OVA as a model allergen [105] and glucose as a reducing sugar in the Maillard reaction promoted by heating. Compared with the controls (native OVA and OVA thermally processed without glucose), AGE-OVA enhanced the activation of OVA-specific CD4(+) T cells on coculture with myeloid dendritic cells (mDCs), indicating that the glycation of OVA enhanced the T-cell immunogenicity of the allergen. The bone-marrow-derived murine myeloid DC uptake of AGE-OVA was significantly higher than that of the controls. Scavenger receptor class A type I and II (SR-AI/II) was identified as a mediator of the AGE-OVA uptake, whereas the receptors for AGEs and galectin-3 were not responsible. The activation of OVA-specific CD4(+) T cells by AGE-OVA was attenuated on coculture with SR-AI/II-deficient mDCs [105]. The Maillard reaction might thus play an important role in the T cell immunogenicity of food allergens [106].

The Cry j 1-galactomannan and Cry j 1-mannose conjugates were effectively trafficked in the gut and co-localized with immune cells, such as dendritic cells in the gut, suggesting that Cry j 1-saccharide conjugates are phagocytosed via the mannose receptor in immune cells. These results suggest that the Cry j 1-galactomannan conjugate is suitable for masking the epitope sites of Cry j 1 and trafficking to immune cells in gut lumen [107].

Conjugation of major buckwheat allergen Fag e 1 with arabinogalactan, xyloglucan, or yeast glucomannan by the Maillard reaction decreased in vitro allergenicity of the protein. Determination of IgE titer in the allergic mice revealed that conjugation with yeast glucomannan was the most effective for decreasing in vivo allergenicity of Fag e 1 among these water-soluble polysaccharides [108].

Rupa et al. studied the various mannose-related glycated forms of OVA (monosaccharides and polysaccharides) and role of these forms in tolerance induction in a Balb/c mouse model. Glycated forms of OVA such as OVA-mannose (OVAMan-a monosaccharide), glucomannan and galactomannan (OVAGluMan, OVAGalMan-polysaccharides), and a mixture containing glucomannan and OVA (mix) were used for treatment modalities. The data clearly indicated that both OVAMan and OVA-GluMan were able to suppress allergic immune response in mice [109]. In agreement with previous studies, glycated forms of OVA and glucose were not able to suppress allergic immune response [105].

Literature data thus show a great variation in the immune response to glycated forms of allergens, which may depend on the type of saccharide used for modification, but could also be allergen specific.

5.3.3.3   Conjugation of Allergens with Polysaccharides

Carboxymethyldextran Hattori et al. [110, 111] prepared β-lactoglobulin-carboxymethyl dextran conjugates (BLG-CMD), by using a water-soluble carbodiimide. The anti-BLG antibody response was markedly reduced after immunization with the BLG-CMD conjugates in mice. Linear epitope profiles of the BLG-CMD conjugates were similar to those of BLG, while the antibody response for each epitope was dramatically reduced. Reduction of immunogenicity of BLG depended on CMD content indicating that masking of epitopes by CMD is responsible for the decreased immunogenicity of the BLG conjugates due to effective shielding by CMD.

Similar results were obtained by Kobayashi et al. [112] who prepared BLG-CMD with different molar ratios. Results of both studies show that conjugation with CMD of higher molecular weight is effective in reducing the immunogenicity of BLG by masking of B-cell epitopes by CMD. In their further study, Kobayashi et al. [113] investigated changes in the T-cell response to BLG after conjugation with CMDs. When lymph node cells from mice immunized with BLG or the conjugates were stimulated with BLG, T cells from the conjugate-immunized mice showed a lower proliferative response comparing to BLG-immunized mice. T-cell epitope profiles of the conjugates were similar to those of BLG, whereas the proliferative response to each epitope was reduced, indicating that the lower in vivo T-cell response with the conjugates was not due to induction of conjugate-specific T cells, but due to a decrease in the number of BLG-specific T cells. In addition, conjugation with CMD enhanced the resistance of BLG to cathepsin B and cathepsin D, suggesting that conjugation with CMD inhibited the degradation of BLG by proteases in APC and led to suppression of the generation of antigenic peptides including T-cell epitopes from BLG. Therefore, the authors considered that the suppressive effect on the generation of T-cell epitopes reduced the antigen presentation of the conjugates and this reduction led to a decrease in the number of BLG-specific T cells in vivo. As a result, the decreased help to B cells by T cells would have reduced the antibody response to BLG leading to the conclusion that suppression of the generation of T-cell epitopes by conjugation with CMD is important to the mechanism for the reduced immunogenicity of BLG.

Acidic polysaccharides Many studies on neoglycoconjugates of proteins have been performed during the past 20 years, and various improvements in the functional properties of proteins have been achieved: improvement of protein solubility, heat stability, foaming properties, and emulsifying properties. Hattori et al. [114] conjugated BLG with the acidic oligosaccharides, alginic acid oligosaccharide (ALGO) and phosphoryl oligosaccharides (Pos), by the Maillard reaction. Fluorescence studies indicated that the surface of each conjugate was covered with a saccharide chain. The anti-BLG antibody response was markedly reduced after immunization with both conjugates in mice. Linear epitope profiles of the conjugates were found to be similar to those of BLG, whereas the antibody response to each epitope was dramatically reduced. In particular, effective reduction of the antibody response was observed in the vicinity of the carbohydrate-binding sites. Obtained conjugates are edible, and have higher thermal stability and improved emulsifying properties than those of native BLG, thus being very useful for food application. Yoshida et al. [115] demonstrated that the T-cell response was reduced when mice were immunized with BLG-ALGO conjugates and that novel epitopes were not generated by conjugation. The authors clarified that the BLG-ALGO conjugate modulated the immune response to Th1 dominance and considered that this property of the BLG-ALGO conjugate would be effective for preventing food allergy as well as by its reduced immunogenicity. Therefore, conjugation with acidic oligosaccharides could be applied to various food allergens to achieve reduced allergenicity with multiple improvements in their properties. The shielding of IgE-binding epitopes on food allergens by materials having low antigenicity and immunogenicity may be an efficient way of reducing allergenicity of the protein, especially with the use of a high-molecular weight modifier to achieve effective shielding of epitopes.

Galactomannan and glucomannan Soy protein–galactomannan conjugate prepared by the Maillard reaction decreased the allergenicity of the 34-kDa protein which is frequently recognized by the IgE antibody in the sera of soybean-sensitive patients as a major allergen [116]. Monitoring of polyclonal antibody titers by an indirect ELISA and immunoblotting of rabbit sera, monoclonal antibody, and human allergic sera showed that soy protein-galactomannan conjugation was more effective in reducing the allergenicity of the soy protein than transglutaminase (TG) treatments and/or chymotrypsin. Additionally, heat stability, solubility, and emulsifying properties were greatly improved by conjugation with galactomannan.

A recent study demonstrated the potential of OVA-glycated glucomannan as a potential beneficial dietary intervention for allergy [109]. Promising data were also obtained for yeast galactomannan-conjugated major buckwheat allergen, Fag e 1 [108]. Conjugation with arabinogalactan, xyloglucan, or yeast glucomannan successfully decreased in vitro allergenicity of Fag e 1.

Determination of IgE titer in the tested allergic mice revealed that yeast glucomannan was the most effective for in vivo allergenicity of Fag e 1 among the tested water-soluble polysaccharides.

Chitosan Aoki et al. [117] conjugated BLG with chitosan (CHS) by means of a water-soluble carbodiimide to reduce its immunogenicity. The antigenicity of the BLG-CHS conjugates was similar to that of BLG in C3H/He mice, while immunogenicity of BLG reduced by conjugation. The linear epitope profiles of the conjugates were found to be similar to those of BLG, while the antibody response to each epitope dramatically reduced. The researchers suggested masking of B-cell epitopes as one of the mechanisms in reduction of immunogenicity.

Dextran-glycylglycine and amylose-glycylglycine Nodake et al. [118] conjugated BLG with the *N*-hydroxysuccinimide ester of the dextran-glycylglycine adduct (DG-ONSu) to reduce the immunogenicity of BLG. Conjugation with DG-ONSu greatly decreased the immunoreactivity of BLG with anti-BLG antibodies and suppressed their production in vivo presumably due to its shielding effect on epitope(s) on the protein's molecular surface. Besides, DG-BLG was resistant to proteolytic enzymes. In another study [119] of the same group of authors, it was demonstrated that conjugation of BLG with the *N*-hydroxysuccinimide ester of the amylose-glycylglycine adduct (AG-ONSu) also greatly decreased the reactivity of BLG. The authors proposed the usage of DG-ONSu and AG-ONSu to suppress the hypersensitivity mediated by IgE antibodies in milk allergy.

5.3.3.4 Reduction and Modifications of Cystein Residues in Allergens

The proteins are allergenically active and less digestible in the oxidized state, when cysteins are bridged into cystins. When reduced (free sulfhydrils are available), they lose their allergenicity and/or become more digestible. Allergen reduction can be performed by using a reducing agent such as 2-mercaptoethanol, dithiothreitol, cysteine, glutathione, etc., or by using proteins glutaredoxin or thioredoxin. In a study by Buchanan et al. [120], thioredoxin mitigated the allergenicity of whey flour proteins, gliadins, and glutenins, as determined by skin tests with a canine model of food allergy, but gave less consistent results with albumins and globulins. In the study by de Val et al. [121] after reduction of one or both of its disulfide bonds by thioredoxin, BLG became strikingly sensitive to pepsin in SGF and lost its allergenicity as determined by skin test responses and gastrointestinal symptoms in inbred colony of high IgE-producing dogs sensitized to milk.

Koppelman et al. [122] reduced 2S albumin of Brazil nut (Ber e 1) by thioredoxin. Disruption of disulfide bonds was followed by alkylation in order to prevent reformation of disulfide bonds. Far-UV CD and infrared spectroscopy showed that the reduced and alkylated form had lost its β-structures, whereas the α-helix content was conserved. Oral administration of native 2S albumin resulted in the development of IgG1, IgG2a and IgE responses in the rat, as determined by ELISA. Oral exposure to reduced and alkylated 2S albumin (RA-2S albumin) did not result in the development of specific IgE against RA-2S, but IgG1 and IgG2a antibodies against RA-2S albumin were formed in a lower level compared to native 2S albumin. Dosing of the animals with the low dose RA-2S albumin (0.1 mg protein/rat/day) did not result in an antibody response at all in the rats, whereas the same dose of native 2S albumin induced specific IgG1, IgG2a, and IgE responses, again indicating a lesser immunogenicity. Taken together, these data show that reduction of the disulfide bonds of 2S albumin results in the loss of allergenicity and in an increased sensitivity to digestion. All these results provide evidence that thioredoxin can be applied to enhance digestibility and lower allergenicity of food proteins. However, thioredoxins represent a novel family of cross-reactive allergens involved in the pathogenesis of atopic eczema and asthma. Also, cross-reactivity to human thioredoxin can contribute to the exacerbation of severe atopic diseases by involvement of IgE-mediated autoreactivity [123]. Considering these facts, usage of thioredoxin in food allergen modification might be limited.

Within the group of isoforms of peanut conglutin, it has been shown that reduction of the disulfides and subsequent alkylation of the resulting sulfhydryl groups leads to diminished IgE binding [124]. Apparently, the IgE binding mainly depends on the protein structure for these allergens. The reduced and alkylated molecules are hypoallergenic, but still immunogenic [124] and potentially suitable for immunotherapy in peanut-allergic patients. Thus, the peanut conglutins Ara h 2 and Ara h 6 can be chemically modified by reduction and alkylation, such that they substantially unfold and that their allergenic potency decreases. Using conditions for limited reduction and alkylation, partially reduced and alkylated proteins were found with rearranged disulfide bridges and, in some cases, intermolecular cross-links were found. Peptide mass finger printing was applied to control progress of the modification reaction and to map novel disulfide bonds. There was no preference for the order in which disulfides were reduced, and disulfide rearrangement occurred in a nonspecific way. Only minor differences in kinetics of reduction and alkylation were found between the different conglutin isoforms [125].

5.3.4   Enzymatic Modifications of Allergens

Most frequently used enzymes in food processing are proteases and cross-linking enzymes. By the action of proteases on proteins substrates, degradation of polypeptide chain occurs, resulting in a mixture of peptides. By contrast, cross-linking enzymes are able to covalently bridge proteins, thus resulting in large aggregates of proteins, with a molecular weight of up to tens of millions of daltons [126].

Enzymatic cross-linking of proteins, by TG, peroxisases, and phenol oxidases (such as tyrosinases and laccases), is currently exploited in the food-processing industry [126]. Cross-linking enzymes have become a very useful bioprocessing tool in the food industry to improve texture and mechanical properties of food [127]. Oxidases, tyrosinases, laccases, and TG are often exploited for that purpose.

A study comparing four different cross-linking enzymes and β-casein allergenicity demonstrated potential of cross-linking enzymes which use phenolic mediators in creation of less allergenic food products. Cross-linking reduced the digestibility of β-casein [128]. Especially the presence of caffeic acid hampered digestion by pepsin, and this effect was most pronounced for the tyrosinase/caffeic acid cross-linked CN. The laccase/caffeic acid and mushroom tyrosinase/caffeic acid had the highest potential in mitigating IgE binding and allergenicity of the β-casein out of all investigated enzymes. The presence of a small phenolic compound also increased digestion stability of β-casein. Other studies that investigated the effects of polyphenols as mediators of food allergens cross-linking also demonstrated that they can be used from natural sources and bring new functionalities to protein [129, 130].

Peanut protein polymers and glycoprotein conjugates created by TG, exhibited similar IgE-binding activity, compared to control solutions. These results suggested that potential allergic responses were not enhanced after enzymatic modification [131]. In addition, a study showed that allergenic properties of roasted peanut may be reduced by peroxidase (POD) treatment, that also led to moderate polymerization of peanut allergens [132]. Most importantly, aggregation of peanut allergens by tyrosinases, did not increase its allergenic potential in vivo [133].

5.3.4.1   Modification of Allergens by Reaction with TGs

TGs catalyze the formation of a covalent bond between a primary amine (including ε-amino group of lysine residues) and the γ-carboxamide group of protein-bound glutamine leading to protein cross-linking.

Villas-Boas et al. [134] polymerized heat-treated BLG and TG (BLG-TG) and untreated BLG in the presence of cysteine and TG (BLG-Cys-TG). BALB/c mice sensitized with BLG-Cys-TG showed lower levels of IgG1 and IgE than those immunized with native BLG or BLG-TG, suggesting that polymerization in the presence of Cys modified or hid epitopes, reducing the potential antigenicity of BLG.

Clare et al. cross-linked peanut flour (PF) dispersions with TG in the presence and absence of the dithiothreitol [135]. TG treatment did not diminish IgE-binding responses in ELISA, implying only that TG cross-linking does not enhance potential for allergic responses. In their further study, Clare et al. [136] cross-linked light-roasted PF with TG with CN as co-substrate. The functionality of light-roasted PF dispersions containing supplemental CN was altered after polymerization with microbial TG (TGase). In immunoblotting, in some patients' sera, IgE binding to TGase-treated PF-CN fractions appeared less compared to equivalent polymeric PF dispersions lacking supplemental CN and non-cross-linked PF-CN samples. The researchers assumed that covalent modification masked IgE peanut protein-binding epitopes, at least to some degree, on an individual patient basis.

In a study by Wroblewska et al. [137], whey protein concentrate (WPC) was modified by two enzymes: proteinase alcalase and TG. The new products were characterized by 2D electrophoresis, immunoblotting, and ELISA methods. The WPC hydrolysate obtained with alcalase contained proteins and peptides showing strong immunoreactive properties, as revealed by immunoblotting with ALA and BLG polyclonal rabbit antibodies. However, the immunoblot analysis demonstrated that WPC showed a stronger reactivity towards IgE of allergic patients than WPC treated with TG. ELISA assay with human sera showed that two-step modification with alcalase followed by TG significantly reduced the immunoreactive properties of whey proteins. Patients with wheat-dependent, exercise-induced anaphylaxis (WDEIA) experience recurrent anaphylactic reactions when exercising after ingestion of wheat products.

Only a few of the numerous wheat proteins recognized by IgE of sensitized individuals have been characterized at the molecular level. Characterized allergens causing baker's asthma include several water/salt-soluble wheat proteins, however sensitization patterns show a great degree of individual variation.

Leszczynska et al. [138] modified wheat flour by the treatment with TG and demonstrated, in indirect noncompetitive ELISA with human sera, reduction of glutenin immunoreactivity to below 30%. However, addition of TG to cereal products can generate epitopes responsible for celiac disease [139]. The insoluble gliadins have been implicated in IgE-mediated allergy to ingested wheat, and omega-5 gliadin has been identified as a major allergen in wheat-dependent, exercise-induced anaphylaxis. The presence of IgE to purified omega-5 gliadin in children was highly predictive of immediate clinical symptoms on oral wheat challenge [140]. Palosuo et al. digested purified ω-5 gliadin, major allergen in WDEIA, with pepsin or with pepsin/trypsin, and treated with tissue transglutaminase (tTG). The IgE-binding ability of ω-5 gliadin was retained after pepsin and pepsin–trypsin digestion, as shown in IgE ELISA test. tTG treatment of the whole peptic digest resulted in cross-linked aggregates which bound IgE antibodies in immunoblotting more intensely than untreated, pepsin-digested, or pepsin–trypsin-digested ω-5 gliadin. In the 20 WDEIA patients, the mean skin prick test wheal elicited by tTG-treated peptic fraction was 77% larger than that elicited by the untreated peptic fraction and 56% larger than that elicited by intact ω-5 gliadin. These results suggest that activation of tTG during exercise in the intestinal mucosa of patients with WDEIA could lead to the formation of large allergen complexes capable of eliciting anaphylactic reactions [141].

Soy protein–galactomannan conjugation was more effective in reducing the allergenicity of the soy protein than TG treatments [116]. In a study by Monogioudi et al. [142], β-casein was cross-linked by TG and demonstrated that enzymatically cross-linked β-casein was stable under acidic conditions and was more resistant to pepsin digestion when compared to the non-cross-linked β-casein. In the study by Stanic et al. [128], TG-treated CN showed no mitigated IgE-binding reactivity compared with the untreated CN in basophil activation test (BAT).

As a microbial TG is included in many food technological processes, safety of the TG itself, as well as safety of the deamidated/cross-linked proteins obtained by the action of this enzyme, should be investigated [143]. In their study, Pedersen et al. [144] investigated the allergenicity of TG from *Streptoverticillium mobaraens* by evaluation of amino acid sequence similarity to known allergens, pepsin resistance, and detection of protein binding to specific serum IgE (RAST), evaluated as recommended by 2001 FAO/WHO Decision Tree, recommended for evaluation of proteins from genetically modified organisms (GMOs). All tests demonstrated that there are no safety concerns with regard to the allergenic potential of tested TG.

5.3.4.2 Modification of Allergens by Reaction with Peroxidases

POD is a heme-containing enzyme, catalyzing the oxidation of a variety of organic compounds by hydrogen peroxide or hydroperoxides. Acting on phenolic compounds, POD generate *o*-quinones, which further react with other phenolics, amino, or sulfhydryl compounds in proteins to form cross-linked products.

In their study, Chung et al. [145] have treated protein extracts from raw and roasted defatted peanut meals with POD in the presence of hydrogen. While POD treatment had no effect on raw peanuts, a significant cross-linking and decrease in the levels of the major allergens, Ara h 1 and Ara h 2, in roasted peanuts were observed in immunoblots and IgE ELISA. The authors suppose that POD induced the cross-linking of mainly Ara h 1 and Ara h 2 from roasted peanuts and that, due to POD treatment, IgE binding was reduced.

Garcia et al. [146] investigated effects of POD and diethyldithiocarbamic acid (DIECA) on IgE binding of Mal d 1, the major apple allergen. In competitive ELISA, IgE binding of Mal d 1 decreased by adding POD. DIECA protected the IgE binding by the allergen, protection being less strong in the presence of exogenous POD.

Weangsripanaval et al. [147] purified and characterized new allergenic protein from the tomato identified as suberization-associated anionic POD. Furthermore, Sanchez Monge et al. [148] purified and characterized allergenic protein from wheat flour identified as seed-specific POD. These facts imply that POD themselves can be allergens, and safety of their use must be assessed.

5.3.4.3 Modification of Allergens by Reaction with Phenol Oxidases

Polyphenol Oxidases (Tyrosinases) Polyphenol oxidases (PPO) or tyrosinases are bifunctional enzymes catalyzing O hydroxylation of monophenols (including protein-bound tyrosine residues) to o-diphenols and subsequent oxidation of o-diphenols to o-quinones [149]. Reactive o-quinones can further undergo nonenzymatic polymerization or react with amino acid residues in proteins.

The influence of thermal processing and nonenzymatic as well as polyphenoloxidase-catalyzed browning reaction on the allergenicity of the major cherry allergen Pru av 1 was investigated in the study by Gruber et al. [150]. Incubation of recombinant Pru av 1, major cherry allergen, with phenol compounds in the presence of tyrosinase led to a decrease in IgE-binding activity of the protein as revealed by enzyme allergosorbent test (EAST) and inhibition assays.

Caffeic acid and epicatechin showed to be the most efficient in decreasing of rPru av 1 IgE-binding activity, followed by catechin and gallic acid, while quercetin and rutin were the least efficient. However, PPO without the addition of a phenolic compound did not display a reduction in IgE binding. The researchers speculated that reactive intermediates formed during enzymatic polyphenol oxidation are responsible for modifying nucleophilic amino acid side chains of proteins, thus inducing an irreversible change in the tertiary structure of the protein and resulting in a loss of conformational epitopes of the allergen.

Peanut extracts treated with and without PPO, PPO/caffeic, and caffeic acid [151] resulted in cross-linking and reduction of the IgE binding in competitive inhibition ELISA of two peanut major allergens, Ara h 1 and Ara h 2. Of the three treatments, PPO/caffeic was the most effective in reducing IgE binding and the allergenic properties of peanut allergens.

Novotna et al. [152] investigated effects of celery juice oxidation by its natural PPO on the reduction of the content of the Api g1, the main celery allergen. Oxidation failed to eliminate the allergenicity of pure celery juice, but oxidation in apple-celery juices reduced the allergenicity of the mixture. However, the BAT showed no reduction in the allergic response to the oxidized juice mixture. Skin testing showed that the prolonged oxidation of juice mixture showed significantly lower reaction, while apple juice stabilized with ascorbic acid did not have effect. Due to the contradictory results in different tests, the method cannot be declared successful or safe, even for mixtures of apple-celery juices. In competitive ELISA, Garcia et al. [146] demonstrated decreased IgE binding of Mal d 1 after enrichment of apple extract with PPO, with the strongest effect in presence of catechin. Antioxidant DIECA protected the IgE binding by the allergen, protection being less strong in the presence of exogenous PPO. Schmitz et al. [153] evaluated the relationship between content of main apple allergen, Mal d 1, and PPO, total phenol content and antioxidative capacity in different apple varieties. Although higher PPO activities and polyphenols contents result in less extractable Mal d 1, higher antioxidative activity can inhibit the interaction between oxidized phenols and Mal d 1, resulting in higher allergenicity (extractable Mal d 1). In the study by Monogioudi et al. [142] β-casein was cross-linked by tyrosinase and the authors demonstrated that enzymatically cross-linked β-casein was stable under acidic conditions and was more resistant to pepsin digestion when compared to non-cross-linked β-casein. In the study by Stanic et al. [128], tyrosinase treated CN showed mitigated IgE binding reactivity, compared with the untreated CN in a basophil activation assay.

Laccases Laccases catalyze oxidation of various phenolic compounds with one electron mechanism generating free radical species. Reactive free radicals can further undergo nonenzymatic polymerization or can react with high redox potential substrate targets, such as amino acid residues in proteins [154].

Tantoush et al. [130] cross-linked BLG by laccase in the presence of sour cherry phenolics. In a basophil activation assay, the allergenicity of the cross-linked protein was shown to decrease in all nine cow's milk-allergic patients, while digestibility of the remaining monomeric BLG in simulated conditions of the GIT increased. In a further study by Tantoush et al., cross-linking BLG by laccase in the presence of apple phenolics (APE) rendered the protein insoluble in the reaction mixture consisting of cross-linked BLG, with a fraction of the BLG remaining monomeric [129]. Enzymatic processing of BLG decreases the bi-phasal gastric-intestinal digestibility of the monomeric and cross-linked protein, thus decreasing its nutritional value.

Stanic et al. [128] cross-linked β-casein by laccase and caffeic acid and demonstrated that cross-linking was not very efficient, leaving mostly monomeric CN modified by caffeic acid. Regardless of that, ability of cross-linked CN to activate basophils was significantly reduced in seven patients and reduced inhibition potential is possibly due to hindering of epitopes by monomer modification. Pepsin digestion of CN cross-linked by laccase by pepsin was hampered.

As enzyme preparations used in food technology are food grade, but often not of the highest purity, they can contain contaminating enzyme activities. These so-called side activities even if present only in trace quantities can have an unpredictable influence on functional properties, nutritional quality, and safety of food implying that effects of contaminating enzymes in used enzyme preparations should be carefully monitored. Stanic et al. [155] found out that in the presence of high-purity commercial laccase and tyrosinase preparations, both variants of BLG (A and B), underwent removal of a peptide from the N-terminus. The truncated forms were more susceptible to digestion by pepsin.

5.3.5   Thermal Treatments: Boiling, Heating, and Roasting

Heating generally decreases protein allergenicity by destroying conformational epitopes. The majority of milk and egg-allergic children tolerate extensively heated (baked with wheat matrix) milk and egg. However, heating may also promote chemical reaction of protein modification with sugars present in the food—the Maillard reaction. It has been shown that in peanut and shrimp, heat-induced Maillard reaction (glycation) may increase allergenicity [103, 156].

The stability of an allergen's IgE-binding capacity towards heating is considered an important characteristic for food allergens. Peanut Ara h 3 and soybean glycinin are relatively stable to thermal denaturation, and upon heating, aggregates were formed. Heating slightly decreased the pepsin digestion rate of both allergens. However, heating did not affect the IgE-binding capacity of the hydrolyzates, as after only 10 min of hydrolysis no IgE binding could be detected any more. Peanut allergen Ara h 1, when digested under equal conditions, still showed IgE binding after 2 h of hydrolysis.

The IgE-binding capacity of legumin allergens from peanuts and soybeans does not withstand peptic digestion. Consequently, these allergens are likely unable to sensitize via the GIT and cause systemic food allergy symptoms. These proteins might thus be less important allergens than was previously assumed [157].

Roasting of peanuts resulted in a significant decrease in protein solubility. At pH 2, the solubility increased dramatically. More extensive resolubilization was observed with amylase treatment. The protein released into solution had a high IgE-binding capacity. While amylase was effective at resolubilizing this material, digestive tract proteases were not. The presence of these insolubilized peanut proteins provides that way a continuous source of allergens to the gastrointestinal mucosal immune system [158].

Mal d 2 showed remarkable stability to proteolysis and thermal treatments. The allergen remained intact after 2 h each of gastric and subsequent duodenal digestion retaining its full IgE-binding capacity. Refolding after cooling was only observed at acidic pH. Mal d 2 maintains its structure in the GIT, a feature essential for sensitizing the mucosal immune system and provoking allergic reactions [56].

Thermal treatment may induce irreversible changes in protein secondary structure. Heating of hen egg allergen OVA to 70 °C has only a minor effect on its secondary structure. However, these minor changes lead to different kinetics and occurrence of fragments after digestion with pepsin. This results in activation of different T-cell subpopulations and changes in both cytokine production and specific antibody formation, which leads to a shift towards Th1 response and ultimately reduces OVA allergenicity [159]. The heated OVA fragments still have the ability to induce allergic symptoms, but these are less pronounced and need longer time to develop.

Heat processing of major fish allergen parvalbumin greatly affected its antibody reactivity. While heating caused a reduction in antibody reactivity to multimeric forms of parvalbumins for most bony fish, a complete loss of reactivity was observed for cartilaginous fish. Molecular analysis demonstrated that parvalbumin cross-reactivity, among fish species, is due to the molecular phylogenetic association of this major fish allergen.

Immunoreactive high molecular weight protein aggregates were formed from cooked protein extracts of tuna, salmon, cod, and flounder [160]. Lep w 1, a major allergen of whiff fish and a calcium-binding beta-parvalbumin, was easily digested using physiological gastric conditions. However, food processing such as cooking could generate dimers that were partially stable towards gastric digestion. It is likely that the observation of stable parvalbumin dimers and the formation of protein aggregates after cooking explain the high allergenicity of this fish [69].

Boiling had little impact on the digestive stability of crab tropomyosin. In contrast, combined ultrasound and boiling, and high-pressure steaming both could accelerate the digestion of TM. Similarly, Western blotting and inhibition ELISA also demonstrated that the reactivity of IgG/IgE binding of tropomyosin that was extracted from processed crab was partially decreased after treating with combined treatment or high pressure steaming [70].

5.3.6   Hydrolysis and Pasteurization of Food Allergens

The action of hydrolytic enzymes has often been exploited in making hypoallergenic food products [84, 161]. The obtained hydrolysates lack original allergen 3D structure that can cause allergic reaction [162]. The safety question is now related to the minimal allergen fragment size that is able to sensitize a patient. It has been shown that many of hypoallergenic formula still have a sensitizing capacity [163].

β-Lactoglobulin shows a high stability against peptic hydrolysis in its native form. However, when raw milk or pasteurized milk was fermented, the rate of peptic digestion of the protein significantly increased (up to 45 % in 2 h) together with decrease of immunochemical response.

The results showed that soluble β-lactoglobulin and α-lactalbumin, but not insoluble CN, were readily transcytosed through enterocytes. Pasteurization caused aggregation of β-lactoglobulin and α-lactalbumin inhibiting uptake by intestinal epithelium. Aggregation redirected allergen uptake to Peyer's patches, which promoted significantly higher Th2-associated antibody and cytokine production in mice than their native counterparts. It appears that triggering of an anaphylactic response requires both a sensitization by aggregates through Peyer's patches and an efficient transfer of soluble protein across the epithelial barrier [164].

5.3.7   Modern Nonthermal Food Processing Methods: High Hydrostatic Pressure, Gamma Irradiation, and Sonication

Recent research has shown that nonthermal food-processing methods, such as high-intensity pressure and sonication, may also effectively alter the digestibility of food proteins and its IgE binding ability [165, 166]. Although mild physicochemical procedures of food processing, both high pressure and ultrasound may alter the structure of globular food proteins. Nonthermal food-processing technologies, such as gamma irradiation [167], pulsed electric field [168], and sonication [169] were also reported to promote modification of food proteins by the Maillard reaction.

The use of high-intensity ultrasound (sonication) has attracted considerable attention due to its potential in the development of novel, relatively mild but targeted, processes to improve the quality and safety of processed foods. Comparing to the numerous food-processing methods, sonication is an effective processing and preservation technology. Sonication can induce both thermal and nonthermal changes in protein structure. Due to the creation of extremely high localized temperatures and pressures, free radicals and other reactive species are formed that readily modify protein side residues [170]. As a result of local reactive species generation, proteins containing disulfide bonds may undergo covalent bonds reformation resulting in misfolded variants of proteins, disulfide-bridged dimers, trimers, and higher oligomerized species [171]. The heating and sonication of globular proteins disrupt some of the forces responsible for the stability of tertiary and/or secondary protein structures. These forces include hydrogen interactions between the polar groups and interactions of nonpolar groups through the surrounding water molecules which form cages around hydrophobic groups.

It has been shown that structural changes in BLG induced by sonication under controlled conditions had a minor influence on allergenic properties and retinol-binding function of the protein. Sonication can modify conformation of the protein, while having only a minor effect on the IgE binding that cannot be regarded as clinically significant. The linear allergenic epitopes of BLG seem to be sonication resistant and more important than conformational epitopes in the tested population of cow's milk-allergic patients [171].

High-intensity ultrasound treatment for 15 min at 0 °C did not have an effect on the allergenicity of Pen a 1, the major shrimp allergen, while after sonication at 50 °C, a significant decrease in allergenicity was observed. By prolonging sonication time for 90 min at 0 °C, the authors observed a slightly increased specific IgE response [172].

High pressure is known to affect the structure of proteins; typically, few hundred MPa pressure can lead to denaturation. That is why several trials have been performed to alter the structure of the allergen proteins by high pressure, in order to reduce its allergenicity. Studies have been performed both on protein solutions and on complex food systems.

Pressure-treated major apple allergen, Mal d 1, showed decrease of the helical content and increase of the beta structure as revealed by CD spectroscopy. Although the secondary structure changes detected by the spectroscopic techniques were mild, authors reported an immunological effect of the pressure treatment, as they observed the decrease of the average wheal area as a function of the pressure of the treatment in a group of tested patients. A total of 200 MPa resulted already in 80 % reduction of the wheal area, but 600 MPa was not enough to eliminate the wheal in the case of all the patients [173]. A more recent study reported no significant change in basophil activation tests of Mal d 1 solution due to treatment by pressures of up to 500 MPa. They also performed Western blot to study the IgE binding of the treated Mal d 1. Skin prick-to-prick tests and double-blind placebo-controlled food challenge tests were done on the apple juice. A rigorous statistical evaluation was performed to prove the effect of pressure treatment (up to 550 MPa). None of these tests showed any significant difference compared to placebo [174].

Conformational change in a food protein can also expose hidden epitopes, resulting in an increase in allergenicity and antigenicity of the modified protein. It has been shown by Kleber et al. that treatment with high hydrostatic pressure enhanced the antigenic response to BLG due to thermally induced unfolding and aggregation of the protein [175].

5.4   Food Matrix and Protein Digestion

Among the conditions required for food proteins to trigger an allergic reaction is their ability to keep the integrity of their allergenic determinants through the GIT. A major characteristic of many food allergens is their resistance to gastric digestion. However, several food allergens have been identified as being sensitive to digestion, while many nonallergenic proteins are as resistant to degradation as allergenic proteins. These observations underline the importance of taking into account the complex factors that play a role in physiological digestion and of evaluating the ability of the fragments generated upon digestion to retain biologically relevant IgE epitopes [50]. Most of the studies dealing with the influence of the gastroduodenal digestion on the potential allergenicity of foods have been carried out on isolated proteins. However, exposure of allergic individuals to pure allergens is rare and, in fact, the stability of proteins to digestion can be altered in the presence of various components that form part of the food matrix, such as biopolymers (polysaccharides, polyphenols) [176, 177], lipids [47, 178], and/or small molecules that may inhibit or activate digestion [179].

The food matrix has thus been suggested to affect the allergenic properties of proteins by providing adjuvant stimuli to the specialized gut mucosal immune system or by protecting them from digestion. It has been shown that purified peanut allergens possess little intrinsic immune-stimulating capacity, in contrast to a whole peanut extract, in animal models for allergenic potential assessment [81]. Endogenous Brazil nut lipids are required for the induction of optimal antibody responses to Ber e 1 in the BALB/c strain mouse. Appropriate antibody-binding sites are present on both natural and recombinant forms of Ber e 1, suggesting that the impact of lipid is at the induction phase, rather than antibody recognition, and is possibly required for efficient antigen presentation [82]. Lipid fraction C from Brazil nut provides an essential adjuvant activity to Ber e 1 sensitization, and invariant natural killer T cells play a critical role in the development of Brazil nut-allergic response [180].

5.4.1   Noncovalent Interactions of Food Matrix Components and Food Allergens

Components of food matrix can interact noncovalently with food allergens giving both soluble and insoluble complexes. Complexation with components of food matrix can lower the level of soluble allergens, thus reducing their allergenic properties and/or influencing digestion in the gastrointestinal system by hindering cutting sites from the action of digestive enzymes, and/or directly inhibiting digestive enzymes. Many in vitro and in vivo studies showed antinutritive properties of polyphenols, especially tannins [181, 182]. On the contrary, the effect of some phenolic compounds, namely, resveratrol, catechin, epigallocatechin-3-gallate, quercetin, and polyphenol-rich beverages (i.e., red wine and green tea) on pepsin activity was quite the opposite. The tested polyphenols and beverages increased the initial velocity of the reaction, affecting the maximum velocity of pepsin on denatured hemoglobin as a substrate, and the activating effect was concentration dependent [183].

Phenolic compounds and phytic acid are known to form soluble and insoluble complexes with proteins [184–186]. For instance, it was shown that multivalent hydrophobic interactions cause compaction of cow's milk caseins (CN) with the polyphenol epigallocatechin in way that individual CN molecules "wrap around" polyphenol [187]. A recent study provided data on increasing binding affinities of oligomerized polyphenols to globular proteins [188]. Similarly, a positive correlation was found between the degree of polyphenol oligomerization and inhibition of elastase due to an increased number of protein interacting groups with the enzyme [189].

It appears that differential binding of polyphenols of common beverages to both protease and its protein substrate may determine the balance between pepsin activation by small phenolic compounds and pepsin inhibition effects of the condensed products. The level of oxidation and the way of food processing can thus be used in tailoring protein food digestion.

5.4.1.1 Phytic Acid

Phytic acid, or phytate, is present in the brans and hulls of most grains, beans, nuts, and seeds. Chung et al. [190, 191] treated peanut extract with phytic acid and demonstrated that phytic acid formed complexes with the major peanut allergens (Ara h 1 and Ara h 2), reduced their solubility in acidic and neutral conditions. A sixfold reduction in IgE binding of the extract was observed after treatment with phytic acid, as measured by competitive inhibition ELISA using a pooled serum from peanut-allergic individuals. A similar result obtained with peanut butter slurry led to the suggestion that phytic acid treatment might reduce the allergenicity of peanut-based products due to reducing their solubility. In another study by the same group of authors, a facilitated IgE binding in vitro was observed by peanut allergens and phytic acid. Apparently, phytic acid was able to stabilize allergen–antibody interactions [192]. However, usage of phytic acid might be limited considering its antinutritive properties due to iron chelation.

5.4.1.2 Phenolic ompounds

Adding phenolics such as caffeic, chlorogenic, and ferulic acids to peanut extracts, liquid peanut butter, and peanut butter slurries precipitated most of the major peanut allergens, Ara h 1 and Ara h 2, and complexation was irreversible [190]. Of the three phenolics, caffeic acid formed the most precipitates with peanut extracts. IgE binding was reduced by approximately 10- to 16-fold as determined by inhibition ELISA. Assuming that the insoluble complexes are not absorbed by the body, the researchers concluded that reducing IgE binding by phenolics is feasible and has a great potential in development of less allergenic liquid peanut-based products.

5.4.1.3   Polysaccharides

Foods are complex multicomponent mixtures that can contain proteins and polysaccharides, in many cases interacting as mixed biopolymers. Such interactions can also form in the stomach after ingestion. It has been shown that the presence of soluble polysaccharides, commonly used in the preparation of a wide range of foods, as stabilizers, thickeners, and emulsifiers, reduces protein digestibility. The increase of mixture viscosity, the interactions between the two types of macromolecules, and the inhibition of enzymatic activity have been pointed out to explain this observation. Several studies reported that the interaction of polysaccharides with gastric and duodenal digests of milk or peanut proteins reduces the IgE binding [193, 194], which was attributed to a masking effect on the reactive epitopes. However, other studies indicated that, depending on the nature of the allergens and their digests, their interactions with polysaccharides could also enhance their contact with IgE [195]. The interaction of the main egg allergens, OVA, and OVM with pectin, gum arabic, and xylan increased their IgE binding and hampered their digestibility. The in vitro duodenal digests of OVA and OVM in the presence of the polysaccharides retained a higher IgE binding, probably as a result of the interaction between the polysaccharides and the peptides derived from protein digestion [195].

   Mouecoucou et al. [80] examined the influence of polysaccharides, i.e., gum arabic, low methylated pectin, and xylan, on the in vitro hydrolysis of peanut protein isolate and the in vitro allergenicity of the digestion products. Peanut protein isolate was hydrolyzed in vitro by pepsin, followed by a trypsin/chymotrypsin mixture in dialysis bags. Hydrolysis by all of the digestive enzymes showed retention of some proteins in the dialysis bags in the presence of gum arabic and xylan. The retentates were recognized by IgG and IgE, but IgE binding of retentate containing xylan was reduced. The immunoreactivity of hydrolysis products in dialysates was considerably reduced by polysaccharides.

   Polovic et al. [58] demonstrated that the addition of apple fruit pectin (1.5 and 3 %) to the purified TLP and major kiwi allergen, Act d 2, protected the allergen from in vitro pepsin digestion. Similarly, in vitro digestion of actinidin, a protease and a major allergen of kiwi fruit, was hampered by apple pectin in both gastric and duodenal fluids [196]. In vivo experiments on healthy nonatopic volunteers have shown that 1 h after ingestion of kiwi fruit in gastric content, intact Act d 2 was still present [58]. In their further work Polovic et al. reported that after in vivo digestion of Act d 2 in the presence of apple pectin in rats, both gastric acidity and specific and total pepsin activity, declined and thus protected 23 % of the ingested allergen from digestion for 90 min in vivo [176]. These results show that although presence of polysaccharides can be effective in masking of IgE epitopes, it also reduces allergen digestibility, enabling higher dosages of the allergen to reach the immune system.

   The in vivo data have shown that oral administration of citrus pectin prevented the induction of oral tolerance to OVA in animals and enhanced the penetration of OVA into the serum. Citrus pectin also enhanced the adhesion and production of cytokines (tumor necrosis factor-α, interferon-γ) in peritoneal macrophages [197].

Overall, numerous studies underlined the importance of the food matrix in the digestibility of food allergens and in their potential ability to trigger an immune response.

References

1. Astwood JD, Leach JN, Fuchs RL: **Stability of food allergens to digestion in vitro**. *Nat Biotechnol* 1996, **14**(10):1269–1273.
2. Yagami T, Haishima Y, Nakamura A, Osuna H, Ikezawa Z: **Digestibility of allergens extracted from natural rubber latex and vegetable foods**. *J Allergy Clin Immunol* 2000, **106**(4):752–762.
3. Fu TT, Abbott UR, Hatzos C: **Digestibility of food allergens and nonallergenic proteins in simulated gastric fluid and simulated intestinal fluid—A comparative study**. *J Agric Food Chem* 2002, **50**(24):7154–7160.
4. Fu TJ: **Digestion stability as a criterion for protein allergenicity assessment**. *Ann N Y Acad Sci* 2002, **964**:99–110.
5. Aalberse RC: **Food allergens**. *Environ Toxicol Pharmacol* 1997, **4**(1–2):55–60.
6. Aalberse RC, Stadler BM: **In silico predictability of allergenicity: from amino acid sequence via 3-D structure to allergenicity**. *Mol Nutr Food Res* 2006, **50**(7):625–627.
7. Aalberse RC: **Structural biology of allergens**. *J Allergy Clin Immunol* 2000, **106**(2):228–238.
8. Mills EN, Jenkins JA, Alcocer MJ, Shewry PR: **Structural, biological, and evolutionary relationships of plant food allergens sensitizing via the gastrointestinal tract**. *Crit Rev Food Sci Nutr* 2004, **44**(5):379–407.
9. Shewry PR, Napier JA, Tatham AS: **Seed storage proteins: structures and biosynthesis**. *Plant Cell* 1995, **7**(7):945–956.
10. Kreis M, Forde BG, Rahman S, Miflin BJ, Shewry PR: **Molecular evolution of the seed storage proteins of barley, rye and wheat**. *J Mol Biol* 1985, **183**(3):499–502.
11. Mills EN, Jenkins J, Marigheto N, Belton PS, Gunning AP, Morris VJ: **Allergens of the cupin superfamily**. *Biochem Soc Trans* 2002, **30**(Pt 6):925–929.
12. Breiteneder H, Radauer C: **A classification of plant food allergens**. *J Allergy Clin Immunol* 2004, **113**(5):821–830; quiz 831.
13. Murtagh GJ, Archer DB, Dumoulin M, Ridout S, Matthews S, Arshad SH, Alcocer MJ: **In vitro stability and immunoreactivity of the native and recombinant plant food 2S albumins Ber e 1 and SFA-8**. *Clin Exp Allergy* 2003, **33**(8):1147–1152.
14. Moreno FJ, Maldonado BM, Wellner N, Mills EN: **Thermostability and in vitro digestibility of a purified major allergen 2S albumin (Ses i 1) from white sesame seeds (Sesamum indicum L.)**. *Biochim Biophys Acta* 2005, **1752**(2):142–153.
15. Diaz-Perales A, Blanco C, Sanchez-Monge R, Varela J, Carrillo T, Salcedo G: **Analysis of avocado allergen (Prs a 1) IgE-binding peptides generated by simulated gastric fluid digestion**. *J Allergy Clin Immunol* 2003, **112**(5):1002–1007.
16. Rodriguez-Perez R, Crespo JF, Rodriguez J, Salcedo G: **Profilin is a relevant melon allergen susceptible to pepsin digestion in patients with oral allergy syndrome**. *J Allergy Clin Immunol* 2003, **111**(3):634–639.
17. Sancho AI, Wangorsch A, Jensen BM, Watson A, Alexeev Y, Johnson PE, Mackie AR, Neubauer A, Reese G, Ballmer-Weber B *et al*: **Responsiveness of the major birch allergen Bet v 1 scaffold to the gastric environment: impact on structure and allergenic activity**. *Mol Nutr Food Res* 2011, **55**(11):1690–1699.
18. Hirschwehr R, Valenta R, Ebner C, Ferreira F, Sperr WR, Valent P, Rohac M, Rumpold H, Scheiner O, Kraft D: **Identification of common allergenic structures in hazel pollen and**

hazelnuts: a possible explanation for sensitivity to hazelnuts in patients allergic to tree pollen. *J Allergy Clin Immunol* 1992, **90**(6 Pt 1):927–936.

19. Schocker F, Luttkopf D, Muller U, Thomas P, Vieths S, Becker WM: **IgE binding to unique hazelnut allergens: identification of non pollen-related and heat-stable hazelnut allergens eliciting severe allergic reactions.** *Eur J Nutr* 2000, **39**(4):172–180.

20. Vieths S, Scheurer S, Ballmer-Weber B: **Current understanding of cross-reactivity of food allergens and pollen.** *Ann N Y Acad Sci* 2002, **964**:47–68.

21. Weber RW: **Patterns of pollen cross-allergenicity.** *J Allergy Clin Immunol* 2003, **112**(2):229–239; quiz 240.

22. Pascal M, Munoz-Cano R, Reina Z, Palacin A, Vilella R, Picado C, Juan M, Sanchez-Lopez J, Rueda M, Salcedo G et al: **Lipid transfer protein syndrome: clinical pattern, cofactor effect and profile of molecular sensitization to plant-foods and pollens.** *Clin Exp Allergy* 2012, **42**(10):1529–1539.

23. Salcedo G, Sanchez-Monge R, Diaz-Perales A, Garcia-Casado G, Barber D: **Plant non-specific lipid transfer proteins as food and pollen allergens.** *Clin Exp Allergy* 2004, **34**(9):1336–1341.

24. Pastorello EA, Pompei C, Pravettoni V, Brenna O, Farioli L, Trambaioli C, Conti A: **Lipid transfer proteins and 2S albumins as allergens.** *Allergy* 2001, **56 Suppl 67**:45–47.

25. Wijesinha-Bettoni R, Alexeev Y, Johnson P, Marsh J, Sancho AI, Abdullah SU, Mackie AR, Shewry PR, Smith LJ, Mills EN: **The structural characteristics of nonspecific lipid transfer proteins explain their resistance to gastroduodenal proteolysis.** *Biochemistry* 2010, **49**(10):2130–2139.

26. Asero R, Mistrello G, Roncarolo D, de Vries SC, Gautier MF, Ciurana CL, Verbeek E, Mohammadi T, Knul-Brettlova V, Akkerdaas JH et al: **Lipid transfer protein: a pan-allergen in plant-derived foods that is highly resistant to pepsin digestion.** *Int Arch Allergy Immunol* 2001, **124**(1–3):67–69.

27. Flinterman AE, Akkerdaas JH, den Hartog Jager CF, Rigby NM, Fernandez-Rivas M, Hoekstra MO, Bruijnzeel-Koomen CA, Knulst AC, van Ree R, Pasmans SG: **Lipid transfer protein-linked hazelnut allergy in children from a non-Mediterranean birch-endemic area.** *J Allergy Clin Immunol* 2008, **121**(2):423–428 e422.

28. Sirvent S, Palomares O, Vereda A, Villalba M, Cuesta-Herranz J, Rodriguez R: **nsLTP and profilin are allergens in mustard seeds: cloning, sequencing and recombinant production of Sin a 3 and Sin a 4.** *Clin Exp Allergy* 2009, **39**(12):1929–1936.

29. Pastorello EA, Pompei C, Pravettoni V, Farioli L, Calamari AM, Scibilia J, Robino AM, Conti A, Iametti S, Fortunato D et al: **Lipid-transfer protein is the major maize allergen maintaining IgE-binding activity after cooking at 100 degrees C, as demonstrated in anaphylactic patients and patients with positive double-blind, placebo-controlled food challenge results.** *J Allergy Clin Immunol* 2003, **112**(4):775–783.

30. Vassilopoulou E, Rigby N, Moreno FJ, Zuidmeer L, Akkerdaas J, Tassios I, Papadopoulos NG, Saxoni-Papageorgiou P, van Ree R, Mills C: **Effect of in vitro gastric and duodenal digestion on the allergenicity of grape lipid transfer protein.** *J Allergy Clin Immunol* 2006, **118**(2):473–480.

31. Asero R, Mistrello G, Roncarolo D, de Vries SC, Gautier MF, Ciurana CL, Verbeek E, Mohammadi T, Knul-Brettlova V, Akkerdaas JH et al: **Lipid transfer protein: a pan-allergen in plant-derived foods that is highly resistant to pepsin digestion.** *Int Arch Allergy Immunol* 2000, **122**(1):20–32.

32. Lindorff-Larsen K, Winther JR: **Surprisingly high stability of barley lipid transfer protein, LTP1, towards denaturant, heat and proteases.** *Febs Lett* 2001, **488**(3):145–148.

33. Cavatorta V, Sforza S, Aquino G, Galaverna G, Dossena A, Pastorello EA, Marchelli R: **In vitro gastrointestinal digestion of the major peach allergen Pru p 3, a lipid transfer protein: molecular characterization of the products and assessment of their IgE binding abilities.** *Mol Nutr Food Res* 2010, **54**(10):1452–1457.

34. Scheurer S, Lauer I, Foetisch K, San Miguel Moncin M, Retzek M, Hartz C, Enrique E, Lidholm J, Cistero-Bahima A, Vieths S: **Strong allergenicity of Pru av 3, the lipid transfer**

protein from cherry, is related to high stability against thermal processing and digestion. *J Allergy Clin Immunol* 2004, **114**(4):900–907.

35. Schimek EM, Zwolfer B, Briza P, Jahn-Schmid B, Vogel L, Vieths S, Ebner C, Bohle B: **Gastrointestinal digestion of Bet v 1-homologous food allergens destroys their mediator-releasing, but not T cell-activating, capacity.** *J Allergy Clin Immunol* 2005, **116**(6):1327–1333.

36. Mittag D, Akkerdaas J, Ballmer-Weber BK, Vogel L, Wensing M, Becker WM, Koppelman SJ, Knulst AC, Helbling A, Hefle SL *et al*: **Ara h 8, a Bet v 1-homologous allergen from peanut, is a major allergen in patients with combined birch pollen and peanut allergy.** *J Allergy Clin Immunol* 2004, **114**(6):1410–1417.

37. Lopez-Torrejon G, Crespo JF, Sanchez-Monge R, Sanchez-Jimenez M, Alvarez J, Rodriguez J, Salcedo G: **Allergenic reactivity of the melon profilin Cuc m 2 and its identification as major allergen.** *Clin Exp Allergy* 2005, **35**(8):1065–1072.

38. Lopez-Torrejon G, Ibanez MD, Ahrazem O, Sanchez-Monge R, Sastre J, Lombardero M, Barber D, Salcedo G: **Isolation, cloning and allergenic reactivity of natural profilin Cit s 2, a major orange allergen.** *Allergy* 2005, **60**(11):1424–1429.

39. Wang Y, Fu TJ, Howard A, Kothary MH, McHugh TH, Zhang Y: **Crystal structure of peanut (Arachis hypogaea) allergen Ara h 5.** *J Agric Food Chem* 2013, **61**(7):1573–1578.

40. Bublin M, Pfister M, Radauer C, Oberhuber C, Bulley S, Dewitt AM, Lidholm J, Reese G, Vieths S, Breiteneder H *et al*: **Component-resolved diagnosis of kiwifruit allergy with purified natural and recombinant kiwifruit allergens.** *J Allergy Clin Immunol* 2010, **125**(3):687–694, 694 e681.

41. Le TM, Bublin M, Breiteneder H, Fernandez-Rivas M, Asero R, Ballmer-Weber B, Barreales L, Bures P, Belohlavkova S, de Blay F *et al*: **Kiwifruit allergy across Europe: clinical manifestation and IgE recognition patterns to kiwifruit allergens.** *J Allergy Clin Immunol* 2013, **131**(1):164–171.

42. Sirvent S, Palomares O, Cuesta-Herranz J, Villalba M, Rodriguez R: **Analysis of the Structural and Immunological Stability of 2S Albumin, Nonspecific Lipid Transfer Protein, and Profilin Allergens from Mustard Seeds.** *J Agric Food Chem* 2012.

43. Moreno FJ, Clemente A: **2S Albumin Storage Proteins: What Makes them Food Allergens?** *Open Biochem J* 2008, **2**:16–28.

44. Sen M, Kopper R, Pons L, Abraham EC, Burks AW, Bannon GA: **Protein structure plays a critical role in peanut allergen stability and may determine immunodominant IgE-binding epitopes.** *J Immunol* 2002, **169**(2):882–887.

45. Suhr M, Wicklein D, Lepp U, Becker WM: **Isolation and characterization of natural Ara h 6: evidence for a further peanut allergen with putative clinical relevance based on resistance to pepsin digestion and heat.** *Mol Nutr Food Res* 2004, **48**(5):390–399.

46. Lehmann K, Schweimer K, Reese G, Randow S, Suhr M, Becker WM, Vieths S, Rosch P: **Structure and stability of 2S albumin-type peanut allergens: implications for the severity of peanut allergic reactions.** *Biochem J* 2006, **395**(3):463–472.

47. Moreno FJ, Mellon FA, Wickham MS, Bottrill AR, Mills EN: **Stability of the major allergen Brazil nut 2S albumin (Ber e 1) to physiologically relevant in vitro gastrointestinal digestion.** *FEBS J* 2005, **272**(2):341–352.

48. Kopper RA, Odum NJ, Sen M, Helm RM, Steve Stanley J, Wesley Burks A: **Peanut protein allergens: gastric digestion is carried out exclusively by pepsin.** *J Allergy Clin Immunol* 2004, **114**(3):614–618.

49. Burks AW, Shin D, Cockrell G, Stanley JS, Helm RM, Bannon GA: **Mapping and mutational analysis of the IgE-binding epitopes on Ara h 1, a legume vicilin protein and a major allergen in peanut hypersensitivity.** *Eur J Biochem* 1997, **245**(2):334–339.

50. Eiwegger T, Rigby N, Mondoulet L, Bernard H, Krauth MT, Boehm A, Dehlink E, Valent P, Wal JM, Mills EN *et al*: **Gastro-duodenal digestion products of the major peanut allergen Ara h 1 retain an allergenic potential.** *Clin Exp Allergy* 2006, **36**(10):1281–1288.

51. Maleki SJ, Kopper RA, Shin DS, Park CW, Compadre CM, Sampson H, Burks AW, Bannon GA: **Structure of the major peanut allergen Ara h 1 may protect IgE-binding epitopes from degradation.** *J Immunol* 2000, **164**(11):5844–5849.

52. Bøgh KL, Nielsen H, Madsen CB, Mills ENC, Rigby N, Eiwegger T, Szépfalusi Z, Roggen EL: **IgE epitopes of intact and digested Ara h 1: A comparative study in humans and rats.** *Mol Immunol* 2012, **51**(3–4):337–346.

53. Breiteneder H: **Thaumatin-like proteins—a new family of pollen and fruit allergens.** *Allergy* 2004, **59**(5):479–481.

54. Ho VS, Wong JH, Ng TB: **A thaumatin-like antifungal protein from the emperor banana.** *Peptides* 2007, **28**(4):760–766.

55. Fuchs HC, Bohle B, Dall'Antonia Y, Radauer C, Hoffmann-Sommergruber K, Mari A, Scheiner O, Keller W, Breiteneder H: **Natural and recombinant molecules of the cherry allergen Pru av 2 show diverse structural and B cell characteristics but similar T cell reactivity.** *Clin Exp Allergy* 2006, **36**(3):359–368.

56. Smole U, Bublin M, Radauer C, Ebner C, Breiteneder H: **Mal d 2, the thaumatin-like allergen from apple, is highly resistant to gastrointestinal digestion and thermal processing.** *Int Arch Allergy Immunol* 2008, **147**(4):289–298.

57. Gavrovic-Jankulovic M, Cirkovic T, Vuckovic O, Atanaskovic-Markovic M, Petersen A, Gojgic G, Burazer L, Jankov RM: **Isolation and biochemical characterization of a thaumatin-like kiwi allergen.** *J Allergy Clin Immunol* 2002, **110**(5):805–810.

58. Polovic N, Blanusa M, Gavrovic-Jankulovic M, Atanaskovic-Markovic M, Burazer L, Jankov R, Cirkovic Velickovic T: **A matrix effect in pectin-rich fruits hampers digestion of allergen by pepsin in vivo and in vitro.** *Clin Exp Allergy* 2007, **37**(5):764–771.

59. Poulsen LK, Hansen TK, Norgaard A, Vestergaard H, Stahl Skov P, Bindslev-Jensen C: **Allergens from fish and egg.** *Allergy* 2001, **56 Suppl 67**:39–42.

60. Martos G, Lopez-Fandino R, Molina E: **Immunoreactivity of hen egg allergens: influence on in vitro gastrointestinal digestion of the presence of other egg white proteins and of egg yolk.** *Food Chem* 2013, **136**(2):775–781.

61. Takagi K, Teshima R, Okunuki H, Itoh S, Kawasaki N, Kawanishi T, Hayakawa T, Kohno Y, Urisu A, Sawada J: **Kinetic analysis of pepsin digestion of chicken egg white ovomucoid and allergenic potential of pepsin fragments.** *Int Arch Allergy Immunol* 2005, **136**(1):23–32.

62. Jarvinen KM, Chatchatee P, Bardina L, Beyer K, Sampson HA: **IgE and IgG binding epitopes on alpha-lactalbumin and beta-lactoglobulin in cow's milk allergy.** *Int Arch Allergy Immunol* 2001, **126**(2):111–118.

63. Lam HY, van Hoffen E, Michelsen A, Guikers K, van der Tas CH, Bruijnzeel-Koomen CA, Knulst AC: **Cow's milk allergy in adults is rare but severe: both casein and whey proteins are involved.** *Clin Exp Allergy* 2008, **38**:995–1002.

64. Hong YH, Creamer LK: **Changed protein structures of bovine beta-lactoglobulin B and alpha-lactalbumin as a consequence of heat treatment.** *Int Dairy J* 2002, **12**(4):345–359.

65. Moreno FJ, Mackie AR, Mills EN: **Phospholipid interactions protect the milk allergen alpha-lactalbumin from proteolysis during in vitro digestion.** *J Agric Food Chem* 2005, **53**(25):9810–9816.

66. Dupont D, Mandalari G, Molle D, Jardin J, Leonil J, Faulks RM, Wickham MS, Clare Mills EN, Mackie AR: **Comparative resistance of food proteins to adult and infant in vitro digestion models.** *Mol Nutr Food Res* 2009.

67. Lim DL, Neo KH, Yi FC, Chua KY, Goh DL, Shek LP, Giam YC, Van Bever HP, Lee BW: **Parvalbumin—the major tropical fish allergen.** *Pediatr Allergy Immunol* 2008, **19**(5):399–407.

68. Saptarshi SR, Sharp MF, Kamath SD, Lopata AL: **Antibody reactivity to the major fish allergen parvalbumin is determined by isoforms and impact of thermal processing.** *Food Chem* 2014, **148**:321–328.

69. Griesmeier U, Bublin M, Radauer C, Vazquez-Cortes S, Ma Y, Fernandez-Rivas M, Breiteneder H: **Physicochemical properties and thermal stability of Lep w 1, the major allergen of whiff.** *Mol Nutr Food Res* 2010, **54**(6):861–869.

70. Yu H-L, Cao M-J, Cai Q-F, Weng W-Y, Su W-J, Liu G-M: **Effects of different processing methods on digestibility of Scylla paramamosain allergen (tropomyosin).** *Food Chem Toxicol* 2011, **49**(4):791–798.

71. Huang YY, Liu GM, Cai QF, Weng WY, Maleki SJ, Su WJ, Cao MJ: **Stability of major allergen tropomyosin and other food proteins of mud crab (Scylla serrata) by in vitro gastrointestinal digestion.** *Food Chem Toxicol* 2010, **48**(5):1196–1201.

72. Leung PSC, Chu KH, Chow WK, Ansari A, Bandea CI, Kwan HS, Nagy SM, Gershwin ME: **Cloning, expression, and primary structure of Metapenaeus ensis tropomyosin, the major heat-stable shrimp allergen.** *J Allergy Clin Immunol* 1994, **94**(5):882–890.

73. Leung PSC, Chen Y-c, Gershwin ME, Wong SH, Kwan HS, Chu KH: **Identification and molecular characterization of Charybdis feriatus tropomyosin, the major crab allergen.** *J Allergy Clin Immunol* 1998, **102**(5):847–852.

74. Liu GM, Cao MJ, Huang YY, Cai QF, Weng WY, Su WJ: **Comparative study of in vitro digestibility of major allergen tropomyosin and other food proteins of Chinese mitten crab (Eriocheir sinensis).** *J Sci Food Agric* 2010, **90**(10):1614–1620.

75. Liu GM, Huang YY, Cai QF, Weng WY, Su WJ, Cao MJ: **Comparative study of in vitro digestibility of major allergen, tropomyosin and other proteins between Grass prawn (Penaeus monodon) and Pacific white shrimp (Litopenaeus vannamei).** *J Sci Food Agric* 2011, **91**(1):163–170.

76. van Beresteijn EC, Meijer RJ, Schmidt DG: **Residual antigenicity of hypoallergenic infant formulas and the occurrence of milk-specific IgE antibodies in patients with clinical allergy.** *J Allergy Clin Immunol* 1995, **96**(3):365–374.

77. Huby RD, Dearman RJ, Kimber I: **Why are some proteins allergens?** *Toxicol Sci* 2000, **55**(2):235–246.

78. Thomas K, Aalbers M, Bannon GA, Bartels M, Dearman RJ, Esdaile DJ, Fu TJ, Glatt CM, Hadfield N, Hatzos C et al: **A multi-laboratory evaluation of a common in vitro pepsin digestion assay protocol used in assessing the safety of novel proteins.** *Regul Toxicol Pharmacol* 2004, **39**(2):87–98.

79. Codex Alimentarius Commission.: **Appendix III, Guideline for the conduct of food safety assessment of foods derivedfrom recombinant-DNA plants and Appendix IV, Annex on the assessment of possible allergenicity.** *Alinorm 03/34: Joint FAO/WHO Food Standard Programme, Codex Alimentarius Commission, Twenty-Fifth Session, Rome, Italy* 2003:47e60.

80. Mouecoucou J, Villaume C, Sanchez C, Mejean L: **Beta-lactoglobulin/polysaccharide interactions during in vitro gastric and pancreatic hydrolysis assessed in dialysis bags of different molecular weight cut-offs.** *Biochim Biophys Acta* 2004, **1670**(2):105–112.

81. van Wijk F, Nierkens S, Hassing I, Feijen M, Koppelman SJ, de Jong GA, Pieters R, Knippels LM: **The effect of the food matrix on in vivo immune responses to purified peanut allergens.** *Toxicol Sci* 2005, **86**(2):333–341.

82. Dearman RJ, Alcocer MJC, Kimber I: **Influence of plant lipids on immune responses in mice to the major Brazil nut allergen Ber e 1.** *Clin Exp Allergy* 2007, **37**(4):582–591.

83. Weangsripanaval T, Moriyama T, Kageura T, Ogawa T, Kawada T: **Dietary fat and an exogenous emulsifier increase the gastrointestinal absorption of a major soybean allergen, Gly m Bd 30K, in mice.** *J Nutr* 2005, **135**(7):1738–1744.

84. Penas E, Restani P, Ballabio C, Prestamo G, Fiocchi A, Gomez R: **Evaluation of the residual antigenicity of dairy whey hydrolysates obtained by combination of enzymatic hydrolysis and high-pressure treatment.** *J Food Prot* 2006, **69**(7):1707–1712.

85. Chicon R, Lopez-Fandino R, Alonso E, Belloque J: **Proteolytic pattern, antigenicity, and serum immunoglobulin E binding of beta-lactoglobulin hydrolysates obtained by pepsin and high-pressure treatments.** *J Dairy Sci* 2008, **91**(3):928–938.

86. Tesaki S, Watanabe J, Tanabe S, Sonoyama K, Fukushi E, Kawabata J, Watanabe M: **An active compound against allergen absorption in hypoallergenic wheat flour produced by enzymatic modification.** *Biosci Biotechnol Biochem* 2002, **66**(9):1930–1935.

87. Cirkovic TD, Bukilica MN, Gavrovic MD, Vujcic ZM, Petrovic S, Jankov RM: **Physicochemical and immunologic characterization of low-molecular-weight allergoids of Dactylis glomerata pollen proteins.** *Allergy* 1999, **54**(2):128–134.

88. Cirkovic T, Gavrovic-Jankulovic M, Prisic S, Jankov RM, Burazer L, Vuckovic O, Sporcic Z, Paranos S: **The influence of a residual group in low-molecular-weight allergoids of Artemisia vulgaris pollen on their allergenicity, IgE- and IgG-binding properties.** *Allergy* 2002, **57**(11):1013–1020.

89. Perovic I, Milovanovic M, Stanic D, Burazer L, Petrovic D, Milcic-Matic N, Gafvelin G, van Hage M, Jankov R, Cirkovic Velickovic T: **Allergenicity and immunogenicity of the major mugwort pollen allergen Art v 1 chemically modified by acetylation.** *Clin Exp Allergy* 2009, **39**(3):435–446.

90. Szymkiewicz A, Jedrychowski L: **Immunoreactivity of Acetylated and Succinylated Pea Proteins.** *Acta Aliment Hung* 2009, **38**(3):329–339.

91. Mistrello G, Brenna O, Roncarolo D, Zanoni D, Gentili M, Falagiani P: **Monomeric chemically modified allergens: immunologic and physicochemical characterization.** *Allergy* 1996, **51**(1):8–15.

92. Gruijthuijsen YK, Grieshuber I, Stocklinger A, Tischler U, Fehrenbach T, Weller MG, Vogel L, Vieths S, Poschl U, Duschl A: **Nitration enhances the allergenic potential of proteins.** *Int Arch Allergy Immunol* 2006, **141**(3):265–275.

93. Untersmayr E, Diesner SC, Oostingh GJ, Selzle K, Pfaller T, Schultz C, Zhang Y, Krishnamurthy D, Starkl P, Knittelfelder R et al: **Nitration of the egg-allergen ovalbumin enhances protein allergenicity but reduces the risk for oral sensitization in a murine model of food allergy.** *PLoS One* 2010, **5**(12):e14210.

94. Patterson R, Suszko IM, Bacal E, Zeiss CR, Kelly JF, Pruzansky JJ: **Reduced allergenicity of high molecular weight ragweed polymers.** *J Allergy Clin Immunol* 1979, **63**(1):47–50.

95. Wurtzen PA, Lund G, Lund L, Lund K, Henmar H: **Chemical modification of birch allergen extract leads to a reduction in allergenicity as well as immunogenicity.** *J Allergy Clin Immunol* 2006, **117**(2):S326–S326.

96. Yang X, Gieni RS, Mosmann TR, HayGlass KT: **Chemically modified antigen preferentially elicits induction of Th1-like cytokine synthesis patterns in vivo.** *J Exp Med* 1993, **178**(1):349–353.

97. de Jongh HH, Robles CL, Timmerman E, Nordlee JA, Lee PW, Baumert JL, Hamilton RG, Taylor SL, Koppelman SJ: **Digestibility and IgE-binding of glycosylated codfish parvalbumin.** *Biomed Res Int* 2013, **2013**:756789.

98. Iwan M, Vissers YM, Fiedorowicz E, Kostyra H, Kostyra E, Savelkoul HF, Wichers HJ: **Impact of Maillard reaction on immunoreactivity and allergenicity of the hazelnut allergen Cor a 11.** *J Agric Food Chem* 2011, **59**(13):7163–7171.

99. Jimenez-Saiz R, Belloque J, Molina E, Lopez-Fandino R: **Human immunoglobulin E (IgE) binding to heated and glycated ovalbumin and ovomucoid before and after in vitro digestion.** *J Agric Food Chem* 2011, **59**(18):10044–10051.

100. Nakamura A, Sasaki F, Watanabe K, Ojima T, Ahn DH, Saeki H: **Changes in allergenicity and digestibility of squid tropomyosin during the Maillard reaction with ribose.** *J Agric Food Chem* 2006, **54**(25):9529–9534.

101. Nakamura A, Watanabe K, Ojima T, Ahn DH, Saeki H: **Effect of maillard reaction on allergenicity of scallop tropomyosin.** *J Agric Food Chem* 2005, **53**(19):7559–7564.

102. Yang ZH, Li C, Li YY, Wang ZH: **Effects of Maillard reaction on allergenicity of buckwheat allergen Fag t 3 during thermal processing.** *J Sci Food Agric* 2013, **93**(6):1510–1515.

103. Maleki SJ, Chung SY, Champagne ET, Raufman JP: **The effects of roasting on the allergenic properties of peanut proteins.** *J Allergy Clin Immunol* 2000, **106**(4):763–768.

104. Chung SY, Champagne ET: **Association of end-product adducts with increased IgE binding of roasted peanuts.** *J Agric Food Chem* 2001, **49**(8):3911–3916.

105. Ilchmann A, Burgdorf S, Scheurer S, Waibler Z, Nagai R, Wellner A, Yamamoto Y, Yamamoto H, Henle T, Kurts C et al: **Glycation of a food allergen by the Maillard reaction enhances its T-cell immunogenicity: role of macrophage scavenger receptor class A type I and II.** *J Allergy Clin Immunol* 2010, **125**(1):175–183 e171–111.

106. Hilmenyuk T, Bellinghausen I, Heydenreich B, Ilchmann A, Toda M, Grabbe S, Saloga J: **Effects of glycation of the model food allergen ovalbumin on antigen uptake and presentation by human dendritic cells.** *Immunology* 2010, **129**(3):437–445.

107. Aoki R, Saito A, Azakami H, Kato A: **Effects of various saccharides on the masking of epitope sites and uptake in the gut of cedar allergen Cry j 1-saccharide conjugates by a naturally occurring Maillard reaction.** *J Agric Food Chem* 2010, **58**(13):7986–7990.

108. Suzuki Y, Kassai M, Hirose T, Katayama S, Nakamura K, Akiyama H, Teshima R, Nakamura S: **Modulation of immunoresponse in BALB/c mice by oral administration of Fag e 1-glucomannan conjugate.** *J Agric Food Chem* 2009, **57**(20):9787–9792.

109. Rupa P, Nakamura S, Katayama S, Mine Y: **Effects of ovalbumin glycoconjugates on alleviation of orally induced egg allergy in mice via dendritic-cell maturation and T-cell activation.** *Mol Nutr Food Res* 2013.

110. Hattori M, Okada Y, Takahashi K: **Functional changes in beta-lactoglobulin upon conjugation with carboxymethyl cyclodextrin.** *J Agric Food Chem* 2000, **48**(9):3789–3794.

111. Hattori M, Nagasawa K, Ohgata K, Sone N, Fukuda A, Matsuda H, Takahashi K: **Reduced immunogenicity of beta-lactoglobulin by conjugation with carboxymethyl dextran.** *Bioconjug Chem* 2000, **11**(1):84–93.

112. Kobayashi K, Hirano A, Ohta A, Yoshida T, Takahashi K, Hattori M: **Reduced immunogenicity of beta-lactoglobulin by conjugation with carboxymethyl dextran differing in molecular weight.** *J Agric Food Chem* 2001, **49**(2):823–831.

113. Kobayashi K, Yoshida T, Takahashi K, Hattori M: **Modulation of the T cell response to beta-lactoglobulin by conjugation with carboxymethyl dextran.** *Bioconjug Chem* 2003, **14**(1):168–176.

114. Hattori M, Miyakawa S, Ohama Y, Kawamura H, Yoshida T, To-o K, Kuriki T, Takahashi K: **Reduced immunogenicity of beta-lactoglobulin by conjugation with acidic oligosaccharides.** *J Agric Food Chem* 2004, **52**(14):4546–4553.

115. Yoshida T, Sasahara Y, Miyakawa S, Hattori M: **Reduced T cell response to beta-lactoglobulin by conjugation with acidic oligosaccharides.** *J Agric Food Chem* 2005, **53**(17):6851–6857.

116. Babiker EE, Hiroyuki A, Matsudomi N, Iwata H, Ogawa T, Bando N, Kato A: **Effect of polysaccharide conjugation or transglutaminase treatment on the allergenicity and functional properties of soy protein.** *J Agric Food Chem* 1998, **46**(3):866–871.

117. Aoki T, Iskandar S, Yoshida T, Takahashi K, Hattori M: **Reduced immunogenicity of beta-lactoglobulin by conjugating with chitosan.** *Biosci Biotechnol Biochem* 2006, **70**(10):2349–2356.

118. Nodake Y, Fukumoto S, Fukasawa M, Sakakibara R, Yamasaki N: **Reduction of the immunogenicity of beta-lactoglobulin from cow's milk by conjugation with a dextran derivative.** *Biosci Biotechnol Biochem* 2010, **74**(4):721–726.

119. Nodake Y, Fukumoto S, Fukasawa M, Yamasaki N, Sakakibara R: **Preparation and immunological characterization of beta-lactoglobulin-amylose conjugate.** *Biosci Biotechnol Biochem* 2011, **75**(1):165–167.

120. Buchanan BB, Adamidi C, Lozano RM, Yee BC, Momma M, Kobrehel K, Ermel R, Frick OL: **Thioredoxin-linked mitigation of allergic responses to wheat.** *Proc Natl Acad Sci U S A* 1997, **94**(10):5372–5377.

121. del Val G, Yee BC, Lozano RM, Buchanan BB, Ermel RW, Lee YM, Frick OL: **Thioredoxin treatment increases digestibility and lowers allergenicity of milk.** *J Allergy Clin Immunol* 1999, **103**(4):690–697.

122. Koppelman SJ, Nieuwenhuizen WF, Gaspari M, Knippels LM, Penninks AH, Knol EF, Hefle SL, de Jongh HH: **Reversible denaturation of Brazil nut 2S albumin (Ber e1) and implication of structural destabilization on digestion by pepsin.** *J Agric Food Chem* 2005, **53**(1):123–131.

123. Glaser AG, Menz G, Kirsch AI, Zeller S, Crameri R, Rhyner C: **Auto- and cross-reactivity to thioredoxin allergens in allergic bronchopulmonary aspergillosis.** *Allergy* 2008, **63**(12):1617–1623.

124. Starkl P, Felix F, Krishnamurthy D, Stremnitzer C, Roth-Walter F, Prickett SR, Voskamp AL, Willensdorfer A, Szalai K, Weichselbaumer M *et al*: **An unfolded variant of the major peanut allergen Ara h 2 with decreased anaphylactic potential.** *Clin Exp Allergy* 2012, **42**(12):1801–1812.

125. Apostolovic D, Luykx D, Warmenhoven H, Verbart D, Stanic-Vucinic D, de Jong GA, Cirkovic Velickovic T, Koppelman SJ: **Reduction and alkylation of peanut allergen isoforms Ara h 2 and Ara h 6; characterization of intermediate- and end products.** *Biochim Biophys Acta* 2013, **1834**(12):2832–2842.

126. Buchert J, Ercili Cura D, Ma H, Gasparetti C, Monogioudi E, Faccio G, Mattinen M, Boer H, Partanen R, Selinheimo E *et al*: **Crosslinking food proteins for improved functionality.** *Annu Rev Food Sci Technol* 2010, **1**:113–138.

127. Smiddy MA, Martin JE, Kelly AL, de Kruif CG, Huppertz T: **Stability of casein micelles cross-linked by transglutaminase.** *J Dairy Sci* 2006, **89**(6):1906–1914.

128. Stanic D, Monogioudi E, Dilek E, Radosavljevic J, Atanaskovic-Markovic M, Vuckovic O, Lannto R, Mattinen M, Buchert J, Cirkovic Velickovic T: **Digestibility and allergenicity assessment of enzymatically crosslinked beta-casein.** *Mol Nutr Food Res* 2010, **54**(9):1273–1284.

129. Tantoush Z, Mihajlović L, Kravić B, Ognjenović J, Jankov RM, Cirkovic Velickovic T, Stanić-Vučinić D: **Digestibility of β-lactoglobulin following cross-linking by Trametes versicolor laccase and apple polyphenols.** *J Serb Chem Soc* 2011, **76**(6):847–855.

130. Tantoush Z, Stanic D, Stojadinovic M, Ognjenovic J, Mihajlovic L, Atanaskovic-Markovic M, Cirkovic Velickovic T: **Digestibility and allergenicity of beta-lactoglobulin following laccase-mediated cross-linking in the presence of sour cherry phenolics.** *Food Chem* 2011, **125**(1):84–91.

131. Clare DA, Gharst G, Sanders TH: **Transglutaminase polymerization of peanut proteins.** *J Agric Food Chem* 2007, **55**(2):432–438.

132. Chung SY, Maleki SJ, Champagne ET: **Allergenic properties of roasted peanut allergens may be reduced by peroxidase.** *J Agric Food Chem* 2004, **52**(14):4541–4545.

133. Radosavljevic J, Nordlund E, Mihajlovic L, Krstic M, Bohn T, Buchert J, Cirkovic Velickovic T, Smit J: **Sensitizing potential of enzymatically cross-linked peanut proteins in a mouse model of peanut allergy.** *Mol Nutr Food Res* 2013. doi: 10.1002/mnfr.201300403

134. Villas-Boas MB, Vieira KP, Trevizan G, Zollner RD, Netto FM: **The effect of transglutaminase-induced polymerization in the presence of cysteine on beta-lactoglobulin antigenicity.** *Int Dairy J* 2010, **20**(6):386–392.

135. Clare DA, Gharst G, Sanders TH: **Transglutaminase polymerization of peanut proteins.** *J Agric Food Chem* 2007, **55**(2):432–438.

136. Clare DA, Gharst G, Maleki SJ, Sanders TH: **Effects of transglutaminase catalysis on the functional and immunoglobulin binding properties of peanut flour dispersions containing casein.** *J Agric Food Chem* 2008, **56**(22):10913–10921.

137. Wroblewska B, Jedrychowski L, Hajos G, Szabo E: **Influence of Alcalase and transglutaminase on immunoreactivity of cow milk whey proteins.** *Czech J Food Sci* 2008, **26**(1):15–23.

138. Leszczynska J, Lacka A, Bryszewska M: **The use of transglutaminase in the reduction of immunoreactivity of wheat flour.** *Food Agric Immunol* 2006, **17**(2):105–113.

139. Gerrard JA, Sutton KH: **Addition of transglutaminase to cereal products may generate the epitope responsible for coeliac disease.** *Trends Food Sci Technol* 2005, **16**(11):510–512.

140. Palosuo K: **Update on wheat hypersensitivity.** *Curr Opin Allergy Clin Immunol* 2003, **3**(3):205–209.

141. Palosuo K, Varjonen E, Nurkkala J, Kalkkinen N, Harvima R, Reunala T, Alenius H: **Transglutaminase-mediated cross-linking of a peptic fraction of omega-5 gliadin enhances IgE reactivity in wheat-dependent, exercise-induced anaphylaxis.** *J Allergy Clin Immunol* 2003, **111**(6):1386–1392.

142. Monogioudi E, Faccio G, Lille M, Poutanen K, Buchert J, Mattinen ML: **Effect of enzymatic cross-linking of beta-casein on proteolysis by pepsin.** *Food Hydrocolloid* 2011, **25**(1):71–81.

143. Malandain H: **Transglutaminases: a meeting point for wheat allergy, celiac disease, and food safety.** *Eur Ann Allergy Clin Immunol* 2005, **37**(10):397–403.

144. Pedersen MH, Hansen TK, Sten E, Seguro K, Ohtsuka T, Morita A, Bindslev-Jensen C, Poulsen LK: **Evaluation of the potential allergenicity of the enzyme microbial transglutaminase using the 2001 FAO/WHO Decision Tree.** *Mol Nutr Food Res* 2004, **48**(6):434–440.

145. Chung SY, Maleki SJ, Champagne ET: **Allergenic properties of roasted peanut allergens may be reduced by peroxidase.** *J Agric Food Chem* 2004, **52**(14):4541–4545.

146. Garcia A, Wichers JH, Wichers HJ: **Decrease of the IgE-binding by Mal d 1, the major apple allergen, by means of polyphenol oxidase and peroxidase treatments.** *Food Chem* 2007, **103**(1):94–100.

147. Weangsripanaval T, Nomura N, Moriyama T, Ohta N, Ogawa T: **Identification of suberization-associated anionic peroxidase as a possible allergenic protein from tomato.** *Biosci Biotechnol Biochem* 2003, **67**(6):1299–1304.

148. SanchezMonge R, GarciaCasado G, LopezOtin C, Armentia A, Salcedo G: **Wheat flour peroxidase is a prominent allergen associated with baker's asthma.** *Clin Exp Allergy* 1997, **27**(10):1130–1137.

149. Koschorreck K, Richter SM, Swierczek A, Beifuss U, Schmid RD, Urlacher VB: **Comparative characterization of four laccases from Trametes versicolor concerning phenolic C-C coupling and oxidation of PAHs.** *Arch Biochem Biophys* 2008.

150. Gruber P, Vieths S, Wangorsch A, Nerkamp J, Hofmann T: **Maillard reaction and enzymatic browning affect the allergenicity of Pru av 1, the major allergen from cherry (Prunus avium).** *J Agric Food Chem* 2004, **52**(12):4002–4007.

151. Chung SY, Kato Y, Champagne ET: **Polyphenol oxidase/caffeic acid may reduce the allergenic properties of peanut allergens.** *J Sci Food Agric* 2005, **85**(15):2631–2637.

152. Novotna P, Setinova I, Heroldova M, Kminkova M, Pruchova J, Strohalm J, Fiedlerova V, Winterova R, Kucera P, Houska M: **Deallergisation Trials of Pure Celery Juice and Apple-Celery Juice Mixture by Oxidation.** *Czech J Food Sci* 2011, **29**(2):190–200.

153. Schmitz-Eiberger M, Matthes A: **Effect of harvest maturity, duration of storage and shelf life of apples on the allergen Mal d 1, polyphenoloxidase activity and polyphenol content.** *Food Chem* 2011, **127**(4):1459–1464.

154. Canfora L, Iamarino G, Rao MA, Gianfreda L: **Oxidative transformation of natural and synthetic phenolic mixtures by Trametes versicolor laccase.** *J Agric Food Chem* 2008, **56**(4):1398–1407.

155. Stanic D, Radosavljevic J, Polovic N, Jadranin M, Popovic M, Vuckovic O, Burazer L, Jankov R, Cirkovic Velickovic T: **Removal of N-terminal peptides from β-lactoglobulin by proteolytic contaminants in a commercial phenol oxidase preparation.** *Int Dairy J* 2009, **19**(12):746–752.

156. Nowak-Wegrzyn A, Fiocchi A: **Rare, medium, or well done? The effect of heating and food matrix on food protein allergenicity.** *Curr Opin Allergy Clin Immunol* 2009, **9**(3):234–237.

157. van Boxtel EL, van den Broek LA, Koppelman SJ, Gruppen H: **Legumin allergens from peanuts and soybeans: effects of denaturation and aggregation on allergenicity.** *Mol Nutr Food Res* 2008, **52**(6):674–682.

158. Kopper RA, Odum NJ, Sen M, Helm RM, Stanley JS, Burks AW: **Peanut protein allergens: the effect of roasting on solubility and allergenicity.** *Int Arch Allergy Immunol* 2005, **136**(1):16–22.

159. Golias J, Schwarzer M, Wallner M, Kverka M, Kozakova H, Srutkova D, Klimesova K, Sotkovsky P, Palova-Jelinkova L, Ferreira F *et al*: **Heat-induced structural changes affect OVA-antigen processing and reduce allergic response in mouse model of food allergy.** *PLoS One* 2012, **7**(5):e37156.

160. Bernhisel-Broadbent J, Scanlon SM, Sampson HA: **Fish hypersensitivity. I. In vitro and oral challenge results in fish-allergic patients.** *J Allergy Clin Immunol* 1992, **89**(3):730–737.

161. De La Barca AM, Wall A, Lopez-Diaz JA: **Allergenicity, trypsin inhibitor activity and nutritive quality of enzymatically modified soy proteins.** *International Journal of Food Science and Nutrition* 2005, **56**(3):203–211.

162. Bernasconi E, Fritsche R, Corthesy B: **Specific effects of denaturation, hydrolysis and exposure to Lactococcus lactis on bovine beta-lactoglobulin transepithelial transport, antigenicity and allergenicity.** *Clin Exp Allergy* 2006, **36**(6):803–814.

163. Niggemann B, Nies H, Renz H, Herz U, Wahn U: **Sensitizing capacity and residual allergenicity of hydrolyzed cow's milk formulae: results from a murine model.** *Int Arch Allergy Immunol* 2001, **125**(4):316–321.

164. Roth-Walter F, Berin MC, Arnaboldi P, Escalante CR, Dahan S, Rauch J, Jensen-Jarolim E, Mayer L: **Pasteurization of milk proteins promotes allergic sensitization by enhancing uptake through Peyer's patches.** *Allergy* 2008, **63**(7):882–890.

165. Ross AIV, Griffiths MW, Mittal GS, Deeth HC: **Combining nonthermal technologies to control foodborne microorganisms.** *Int J Food Microbiol* 2003, **89**(2–3):125–138.

166. Husband FA, Aldick T, Van der Plancken I, Grauwet T, Hendrickx M, Skypala I, Mackie AR: **High-pressure treatment reduces the immunoreactivity of the major allergens in apple and celeriac.** *Mol Nutr Food Res* 2011, **55**(7):1087–1095.

167. Chawla SP, Chander R, Sharma A: **Antioxidant properties of Maillard reaction products obtained by gamma-irradiation of whey proteins.** *Food Chem* 2009, **116**(1):122–128.

168. Guan YG, Lin H, Han Z, Wang J, Yu SJ, Zeng XA, Liu YY, Xu CH, Sun WW: **Effects of pulsed electric field treatment on a bovine serum albumin-dextran model system, a means of promoting the Maillard reaction.** *Food Chem* 2010, **123**(2):275–280.

169. Stanic-Vucinic D, Prodic I, Apostolovic D, Nikolic M, Cirkovic Velickovic T: **Structure and antioxidant activity of beta-lactoglobulin-glycoconjugates obtained by high-intensity-ultrasound-induced Maillard reaction in aqueous model systems under neutral conditions.** *Food Chem* 2013, **138**(1):590–599.

170. Riesz P, Kondo T: **Free radical formation induced by ultrasound and its biological implications.** *Free Radic Biol Med* 1992, **13**(3):247–270.

171. Stanic-Vucinic D, Stojadinovic M, Atanaskovic-Markovic M, Ognjenovic J, Gronlund H, van Hage M, Lantto R, Sancho AI, Cirkovic Velickovic T: **Structural changes and allergenic properties of beta-lactoglobulin upon exposure to high-intensity ultrasound.** *Mol Nutr Food Res* 2012, **56**(12):1894–1905.

172. Li ZX, Linhong CL, Jamil K: **Reduction of allergenic properties of shrimp (Penaeus Vannamei) allergens by high intensity ultrasound.** *Eur Food Res Technol* 2006, **223**(5):639–644.

173. Meyer-Pittroff R, Behrendt H, Ring J: **Specific immuno-modulation and therapy by means of high pressure treated allergens.** *High Pressure Res* 2007, **27**(1):63–67.

174. Houska M, Heroldova M, Vavrova H, Kucera P, Setinova I, Havranova M, Honzova S, Strohalm J, Kminkova M, Proskova A et al: **Is high-pressure treatment able to modify the allergenicity of the main apple juice allergen, Mal d1?** *High Pressure Res* 2009, **29**(1):14–22.

175. Kleber N, Krause I, Illgner S, Hinrichs J: **The antigenic response of beta-lactoglobulin is modulated by thermally induced aggregation.** *Eur Food Res Technol* 2004, **219**(2):105–110.

176. Polovic N, Obradovic A, Spasic M, Plecas-Solarovic B, Gavrovic-Jankulovic M, Cirkovic Velickovic T: **In vivo digestion of a thaumatin-like kiwifruit protein in rats.** *Food Digestion* 2010, **1**(1–2):5–13.

177. Stojadinovic M, Radosavljevic J, Ognjenovic J, Vesic J, Prodic I, Stanic-Vucinic D, Cirkovic Velickovic T: **Binding affinity between dietary polyphenols and beta-lactoglobulin negatively correlates with the protein susceptibility to digestion and total antioxidant activity of complexes formed.** *Food Chem* 2013, **136**(3–4):1263–1271.

178. Macierzanka A, Sancho AI, Mills ENC, Rigby NM, Mackie AR: **Emulsification alters simulated gastrointestinal proteolysis of beta-casein and beta-lactoglobulin.** *Soft Matter* 2009, **5**(3):538–550.
179. Maleki SJ, Viquez O, Jacks T, Dodo H, Champagne ET, Chung S-Y, Landry SJ: **The major peanut allergen, Ara h 2, functions as a trypsin inhibitor, and roasting enhances this function.** *J Allergy Clin Immunol* 2003, **112**(1):190–195.
180. Mirotti L, Florsheim E, Rundqvist L, Larsson G, Spinozzi F, Leite-De-Moraes M, Russo M, Alcocer M: **Lipids are required for the development of Brazil nut allergy: The role of mouse and human iNKT cells.** *Allergy* 2013, **68**(1):74–83.
181. Bravo L: **Polyphenols: chemistry, dietary sources, metabolism, and nutritional significance.** *Nutr Rev* 1998, **56**(11):317–333.
182. Goncalves R, Mateus N, Pianet I, Laguerre M, de Freitas V: **Mechanisms of Tannin-Induced Trypsin Inhibition: A Molecular Approach.** *Langmuir* 2011, **27**(21):13122–13129.
183. Tagliazucchi D, Verzelloni E, Conte A: **Effect of some phenolic compounds and beverages on pepsin activity during simulated gastric digestion.** *J Agric Food Chem* 2005, **53**(22):8706–8713.
184. Xiao JB, Zhao YR, Wang H, Yuan YM, Yang F, Zhang C, Yamamoto K: **Noncovalent Interaction of Dietary Polyphenols with Common Human Plasma Proteins.** *J Agric Food Chem* 2011, **59**(19):10747–10754.
185. Xiao JB, Mao FF, Yang F, Zhao YL, Zhang C, Yamamoto K: **Interaction of dietary polyphenols with bovine milk proteins: Molecular structure-affinity relationship and influencing bioactivity aspects.** *Mol Nutr Food Res* 2011, **55**(11):1637–1645.
186. Cao H, Shi Y, Chen X: **Advances on the interaction between tea catechins and plasma proteins: structure-affinity relationship, influence on antioxidant activity, and molecular docking aspects.** *Curr Drug Metab* 2013, **14**(4):446–450.
187. Jobstl E, Howse JR, Fairclough JPA, Williamson MP: **Noncovalent cross-linking of casein by epigallocatechin gallate characterized by single molecule force microscopy.** *J Agric Food Chem* 2006, **54**(12):4077–4081.
188. Prigent SVE, Voragen AGJ, van Koningsveld GA, Baron A, Renard C, Gruppen H: **Interactions between globular proteins and procyanidins of different degrees of polymerization.** *J Dairy Sci* 2009, **92**(12):5843–5853.
189. Bras NF, Goncalves R, Mateus N, Fernandes PA, Ramos MJ, Do Freitas V: **Inhibition of Pancreatic Elastase by Polyphenolic Compounds.** *J Agric Food Chem* 2010, **58**(19):10668–10676.
190. Chung SY, Champagne ET: **Reducing the allergenic capacity of peanut extracts and liquid peanut butter by phenolic compounds.** *Food Chem* 2009, **115**(4):1345–1349.
191. Chung SY, Champagne ET: Effects of Phytic acid on peanut allergens and allergenic properties of extracts. *J Agric Food Chem* 2007, 55(22):9054–9058.
192. Chung S, Champagne ET: **Effect of phytic acid on IgE binding to peanut allergens.** *J Allergy Clin Immunol* 2006, **117**(2):S38–S38.
193. Mouécoucou J, Frémont S, Villaume C, Sanchez C, Méjean L: **Polysaccharides reduce in vitro IgG/IgE-binding of β-lactoglobulin after hydrolysis.** *Food Chem* 2007, **104**(3):1242–1249.
194. Mouecoucou J, Fremont S, Sanchez C, Villaume C, Mejean L: **In vitro allergenicity of peanut after hydrolysis in the presence of polysaccharides.** *Clin Exp Allergy* 2004, **34**(9):1429–1437.
195. Jiménez-Saiz R, López-Expósito I, Molina E, López-Fandiño R: **IgE-binding and in vitro gastrointestinal digestibility of egg allergens in the presence of polysaccharides.** *Food Hydrocolloid* 2013, **30**(2):597–605.
196. Polovic ND, Pjanovic RV, Burazer LM, Velickovic SJ, Jankov RM, Cirkovic Velickovic TD: **Acid-formed pectin gel delays major incomplete kiwi fruit allergen Act c 1 proteolysis in in vitro gastrointestinal digestion.** *J Sci FoodAgric* 2009, **89**(1):8–14.
197. Khramova DS, Popov SV, Golovchenko VV, Vityazev FV, Paderin NM, Ovodov YS: **Abrogation of the oral tolerance to ovalbumin in mice by citrus pectin.** *Nutrition* 2009, **25**(2):226–232.

Chapter 6
Microbiota and Allergic Disease

Contents

Abbreviations

APCs	Antigen-presenting cells
CLRs	C-type lectin receptors
DCs	Dendritic cells
DNA	Deoxyribonucleic acid
GALT	Gut-associated lymphoid tissue
GI	Gastrointestinal
IFN	Interferon
IgA	Immunoglobulin A
IL-4	Interleukin-4
IL-10	Interleukin 10
MAMPs	Microorganism-associated molecular patterns
NF-κB	Nuclear factor-κB
NLRs	NOD-like receptors
PRRs	Pattern recognition receptors
TGF	Transforming growth factor
Th	T helper
TLRs	Toll-like receptors

Summary Probiotics and prebiotics are increasingly being added to foodstuffs with claims of health benefits for humans. Probiotics are live microorganisms which when administered in adequate amounts confer a health benefit on the consumer, whereas prebiotics are ingredients that stimulate the growth and/or function of

T. Ćirković Veličković, M. Gavrović-Jankulović, *Food Allergens,*
Food Microbiology and Food Safety, DOI 10.1007/978-1-4939-0841-7_6,
© Springer Science+Business Media New York 2014

beneficial intestinal microorganisms. The maintenance of intestinal immune and metabolic homeostasis in mammals is strongly affected by the interactions between the mucosa and the intestinal microbiota. Modulation of the intestinal microbiota with probiotics or prebiotics may provide a means for improving the immune status in humans. Selected probiotics have been shown to modulate the immune response, while for prebiotics this has not been fully elucidated. Prebiotics and probiotics could provide safe means of improving the immunological development and functioning. However, to improve the application of probiotics to support and stimulate human health, it is important to improve our understanding of their mode of action.

6.1 The Intestinal Microbiota

Some predictions state that by 2050, two thirds of the global human population will live in urban areas with little green space and with limited contact with nature. At the same time, an increasing number of the urban population will suffer from chronic inflammatory disorders [1, 2], of which allergic [3] and autoimmune diseases are prime examples. Building on the hygiene hypothesis [4, 5], the notion that growing up in a farming environment protects children from allergic sensitization [6], and the emerging understanding of the role of microbes in the development and maintenance of epithelial cell integrity and tolerance [2, 7], the "biodiversity hypothesis" [8] proposed that reduced contact of people with natural environmental features and biodiversity, including environmental microbiota, leads to inadequate stimulation of immunoregulatory circuits [9].

The intestinal microbiota is of great importance to human health and well-being. Modulation of the intestinal microbiota by exogenous and endogenous substrates can be expected to improve various physiological functions of our body. The normal human intestinal microbiota has a diverse composition with more than 1,000 species [10]. This microbiota has a metabolic activity that equals that of the liver, metabolically the most active organ. The microbiota contributes to the digestion of exogenous and endogenous substrates, such as fibers and mucins [10]. This provides the host with additional energy in the form of fatty acids [11]. Another important function of the intestinal microbiota is to provide a protective barrier against bacteria, including potential pathogens, and provide stimulation of the immune system. Several probiotic strains have been observed to modulate some aspects of the immune system; however, in many cases, it is uncertain what the actual health benefit of this immune modulation is. It also remains unclear whether the immunomodulatory effects of probiotics are short term or are sustained and/or reproducible.

In vivo studies in healthy human volunteers measured the changes in gene transcription profiles to determine the molecular responses that occur in the human duodenal mucosa following consumption of probiotic *Lactobacillus* sp. [12]. These studies showed that the mucosal responses to distinct lactobacilli are profoundly different, illustrating the specificity of the host responses to specific bacterial strains and/or species or even different preparations of the same bacterial strain.

The human intestine performs many diverse functions including digestion of food and absorption of nutrients, and it is the largest immune organ of the human body, containing high amounts of antibody-producing cells [13]. The intestinal microbiota is also an important part of the intestinal mucosal barrier. It is therefore not surprising that the intestinal microbiota and the intestinal immune system influence each other and together have an influence on the host beyond the intestine. The immune system regulates the colonization of the intestinal microbiota by interfering with its ability to bind to the mucosa, while parts of bacterial cells and metabolites modulate the immune systems activity [14]. The gastrointestinal (GI) tract provides a wide range of environments for food digestion, varying in pH, flow rate, nutrient availability, etc. Because of these differences, the composition of the microbiota varies throughout the GI tract.

The oral cavity provides different habitats and is colonized by a wide range of organisms. During the first few months of life, only mucosal surfaces and the tongue exist as colonization sites. The oral cavity may be colonized by 500 different bacterial species [15], including streptococci, *Veillonella, Neisseria,* and *Actinomyces,* as the most common genera, depending on the habitat.

The stomach is characterized by acidic pH that may be as low as 1 or 2 [15, 16]; however, in new born infants, it may be close to neutral. The low pH restricts the level of colonization, and the microorganisms found are usually aciduric species such as lactobacilli, streptococci, and *Candida albicans*. In addition, a high percentage of people are colonized by *Helicobacter pylori*. Its natural habitat appears to be the mucus-covered nonacid-secreting epithelium of the antrum. Many factors contribute to the induction of *H. pylori* cells to change to become pathogenic after many years of being a commensal [17]. The duodenum has a sparse microbiota due to the low pH of the digesta released from the stomach and the secretion of pancreatic juice and bile. Also, the swift flow of the digesta reduces the chance for colonization [15]. The composition of the microbiota resembles that of the stomach. In the jejunum, the normal microbiota consists of streptococci, lactobacilli, *Haemophilus, Veillonella, Bacteroides, Corynebacterium,* and *Actinomyces* [18]. Due to the slower passage of the digesta in the ileum, the composition of the microbiota resembles that of the colon with facultative anaerobic Enterobacteriaceae and obligate anaerobes such as *Bacteroides, Veillonella, Clostridium,* lactobacilli, and enterococci are also present [15].

The colon contains the highest density and diversity of microorganisms in the body. The major genera in the colon are *Bacteroides, Bifidobacterium, Clostridium, Eubacterium, Bacillus, Peptostreptococcus, Fusobacterium,* and *Ruminococcus* [15].

6.1.1 Development of Microbiota upon Birth

Upon birth, the intestine is sterile, but it soon becomes colonized by microorganisms from the environment and the mother's birth canal. Bacteria start to appear in the feces within hours after birth, and their numbers increase progressively during the first week of life. The first microorganisms to be isolated from the feces of newborn infants are usually facultative anaerobic organisms such as *Escherichia coli* and other enterobacteria, staphylococci, and streptococci. These organisms change the initially aerobic GI tract to an anaerobic environment which is suitable for colonization by obligate anaerobic organisms [14]. Once the intestine has become anaerobic, bifidobacteria, clostridia, and *Bacteroides* spp. appear in the feces. The fecal colonization of infants born by caesarean section has been found to be delayed compared to vaginally born infants. Also, the composition of the fecal microbiota was different after caesarean section [19, 20]. The traditional view holds that breast-fed infants are colonized mainly by bifidobacteria, while formula-fed infants have a mixed microbiota not particularly high in bifidobacteria. However, in recent reports on the composition of the microbiota of infants, such a difference could often not be observed [14]. This has been attributed to an improved composition of infant formulae, current highly hygienic obstetric practices, and improved bacteriological methodologies [21]. At 2 years of age, the microbiota resembles that of an adult [20]. The environment in which people live and the food they consume have also been observed to influence the composition of the intestinal microbiota [22].

6.2   The Intestinal Mucosa

The intestinal immune system is the largest immune organ of the body. The innate and adaptive immune systems in the intestine are closely integrated with the other functions of the intestine, such as the absorption of nutrients. The intestinal mucosa is composed of a one-cell-thick upper layer, the epithelium, separating the highly colonized intestinal lumen from the second, underlying layer, the *lamina propria*. The *lamina propria* is a special type of connective tissue that is virtually sterile and contains various immune cells [23]. The epithelial layer has a bimodal function, maximizing nutrient absorption while preventing the passage of "nonself" luminal components such as bacteria and food components, which would otherwise induce pro-inflammatory host response [24]. Paneth cells and goblet cells in the epithelium contribute to innate immune defenses that support epithelial barrier function [25]. Paneth cells produce an array of antimicrobials (defensins and lysozyme) that prevent close contact of microorganisms with the proliferative cells in the crypts, whereas goblet cells produce mucins that form a protective layer on the epithelium and prevent direct epithelial contact with luminal microorganisms. The intestinal epithelial barrier, together with innate immune defenses and a mucus layer saturated with antimicrobial peptides, reduces the bacterial load at the interface between the lumen and epithelium [23].

Adaptive immunity in the gut mucosa is largely contained by the *gut-associated lymphoid tissue* (GALT) of the *lamina propria,* such as the Peyer's patches in the small intestine. These patches contain follicle centers and are covered by a specialized follicle-associated epithelium containing microfold cells (M cells), which form a portal for antigen entry into the dome area of the follicle [25]. The mucosa between the Peyer's patches and follicles consists of stromal connective tissue and the immune cells, predominantly B cells, macrophages, dendritic cells (DCs), and T cells. For probiotic functions, the most important classes of T cell in the *lamina propria* are the Th and regulatory T (Treg) cells [26]. Treg cells constitute a major source of the anti-inflammatory interleukin-10 (IL-10) and are involved in the regulation of appropriate T-cell-mediated immune responses, suppression of autoreactive T cells, and maintenance of immune tolerance [27]. Macrophages and DCs in the *lamina propria* of the large and small intestines are dedicated phagocytes. Macrophages are mainly involved in the removal of cellular debris and pathogens, but can also act as antigen-presenting cells (APCs) to initiate adaptive immunity [28], whereas DCs are specialized APCs that regulate both adaptive and innate immunity, for example, by induction of Treg cell development in the presence of retinoic acid (Fig. 6.1). DCs are found throughout the *lamina propria* and Peyer's patches in an immature state that can activate the nuclear factor-κB (NF-κB) pathway, leading to DC maturation and activation [29] on exposure to so-called microorganism-associated molecular patterns (MAMPs) or other stimuli. Activated DCs produce costimulatory molecules, including transforming growth factor-β (TGF-G), retinoic, acid and cytokines such as IL-10, in ratios that depend on the mode of DC stimulation (in this case, MAMP exposure); DCs can also modulate the activation, clonal expansion, and differentiation of T cells [30, 31].

The interaction between epithelium, macrophages, and DCs in response to the luminal microbiota, and the resulting T-cell differentiation and ratios, leads to intestinal immune homeostasis [29, 32]. Homeostasis is also regulated by various pattern recognition receptors (PRRs), which recognize MAMPs derived from bacteria, including probiotics. PRRs include Toll-like receptors (TLRs), nucleotide-binding oligomerization domain (NOD)-like receptors (NLRs), and C-type lectin receptors (CLRs) [33–35]. Intestinal homeostasis can be disturbed by the absence or over-representation of certain bacterial groups (also known as dysbiosis). This can lead to changes in the collective microbial metabolism and an altered presentation of microbial factors and patterns to the human immune system. The link between dysbiosis and deviating abundance of certain microbial consortia is further substantiated by the insight that, at least under certain conditions, probiotic supplementation can correct dysbiosis and restore intestinal homeostasis [23, 36].

The principal antibody in the intestine is immunoglobulin A (IgA), which usually is present in a dimeric form. IgA is well suited for its function in the intestine. It is relatively resistant against proteolysis, in particular IgA2, which is important, considering the environment in the intestine. In contrast to IgG, the major systemic immunoglobulin, IgA does not elicit an inflammatory reaction. IgA can thus bind antigens and exclude them from the intestinal mucosa without causing inflammation [13]. The predominant site of antigen sampling in the intestine is the Peyer's patches, in which M cells specifically sample the contents of the gut and transfer antigens to APCs, which than present the antigen to B and T cells. Naive T cells can develop into Th1 or Th2 phenotype. Th1 cells will direct the differentiation of B cells into IgA-producing cells, while Th2 cells direct B cell differentiation towards IgE-producing cells. Interestingly, M cells have a preference for the uptake of IgA-complexed antigens, thus further stimulating the production of IgA. The

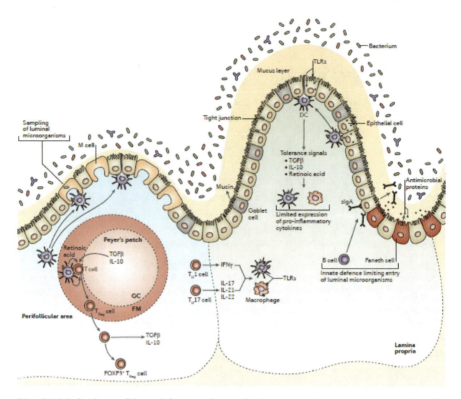

Fig. 6.1 Mechanisms of host defense against and tolerance to intestinal microorganisms. The intestinal environment modulates cellular differentiation in the immune system to control defense against pathogens and tolerance to commensal species. Tolerance depends, among other things, on appropriate innate defense mechanisms that limit microbial entry into intestinal tissues. Intestinal epithelial cells provide a physical barrier between the luminal microorganisms and the underlying intestinal tissues to control homeostasis and tolerance. Specialized epithelial cells produce a mucus layer (*goblet cells*) and secrete antimicrobial proteins (*Paneth cells*) that limit bacterial exposure to the epithelial cells. Production of large amounts of secretory immunoglobulin A (*sIgA*) by B cells provides additional protection from the luminal microbiota. Innate microbial sensing by epithelial cells, dendritic cells (*DCs*), and macrophages is mediated through pattern recognition receptors (*PRRs*) such as Toll-like receptors (*TLRs*). Activation of PRRs on innate immune cells normally induces pathways that mediate microbial killing and activate pro-inflammatory T helper 1 (*Th1*) and Th17 cells and adaptive immune cells. During the maintenance of homeostasis and immune tolerance, however, activation of PRRs on macrophages and DCs in the intestinal lamina propria does not result in secretion of pro-inflammatory cytokines. DCs instead present antigen to T cells in the Peyer's patches and mesenteric lymph nodes, and this can lead to differentiation of regulatory T (*Treg*) cell populations that are regulated by interleukin-10 (*IL-10*), transforming growth factor-β (*TGF-β*), and retinoic acid. Factors secreted by epithelial cells in the intestinal environment can contribute to tolerance of intestinal immune cells. *FM*, follicular mantle; *GC*, germinal centre; *IFNγ*, interferon-γ; *M cell*, microfold cell. (Figure reprinted by permission from Macmillan Publishers Ltd from Ref. [34])

major functions of the intestinal immune system are exclusion of antigens and provision of tolerance to antigens, since all food components and the normal intestinal microbiota are in principle antigens [14]. The oral tolerance is provided through the suppression of Th1 cells by IL-10 and TGF-β when exposed to low concentrations of antigens. High doses cause clonal anergy; T cells are in a state of cellular unresponsiveness which makes them incapable of secreting IL-2 or proliferating [37].

6.3 Microbiota, the Intestinal Immune System, and Allergy

The relevance of the intestinal microbiota for the development of the immune system comes from studies with germfree animals. In the absence of microbes, a mammal has a reduced number of Peyer's patches and less than one tenth of the number of IgA-producing B cells when compared with a conventional animal [38]. Upon exposure to a normal microbiota, ex-germfree animals develop an immune system very much like conventional animals. This indicates the importance of the intestinal microbiota for the development of the immune system. Infants are also born germfree, and the acquisition of the normal microbiota plays an important role in the development of the immune system and the presence of an unbalanced microbiota is associated with disease state [14].

Over the past few decades, an increase in the prevalence of allergy has been observed in industrialized countries [39]. It has been hypothesized that this increase relates to a reduced exposure to microbial antigens as a consequence of increased hygiene and vaccination. This causes a reduced stimulation of the intestinal immune system with bacterial antigens which stimulate the production of Th1 cytokines IL-6, IL-12, IL-18, and IFN-γ, directing the immune system from Th2-mediated immune response [14].

At birth, the immune system of an infant is not fully developed and tends to be directed towards a Th2 phenotype to prevent rejection in utero. The Th2 phenotype leads to the stimulated production of IgE by B cells, and thus increases the risk for allergic reactions through activation of tissue mass cells. Microbial stimulation early in life will reverse the Th2 bias and stimulate the development of a Th1 phenotype and stimulate the activity of Th3 cells [40]. Their combined action will lead to the production of IgA by effector B cells, which contribute to allergen exclusion, and in that way, reduce exposure of the immune system to the allergens.

The increase in allergic diseases has been linked to the relative lack of microbial stimulation, especially in early childhood when the permeability of the gut is higher, and the gut immune system is not fully developed. The immune system of the gut is complexly stimulated by the gut microbiome, which is considered as essential in the evaluation of the hygiene hypothesis, now rephrased as the microbiota hypothesis of allergic diseases [41].

Although the exact etiology of allergic disease is still not clear, many investigators have proposed that environmental exposures may be the major trigger factors in the development of allergy. A low diversity of gut microbiota during the first months of life has been associated with the development of eczema. As the rise in prevalence of allergic diseases has been seen mostly in industrialized countries, this resulted in the definition of the hygiene hypothesis in an attempt to explain the pathogenesis of the disorder [42]. This hypothesis entails that reduced family size and childhood infections have lowered our exposure to microbes, which play a critical role in the maturation of the host immune system during the first years of life [43]. Besides environmental factors, the intestinal microbiota may be a contributor to allergic disease due to its substantial effect on mucosal immunity. Sentinel cells including epithelial cells, macrophages, and intraepithelial dendritic cells continuously sense the environment and coordinate the mucosal immunity. Allergic responses are thought to arise if there is absence of microbial exposure while the immune system is still developing [10, 44]. Exposure to microorganisms early in life allows for a change in the Th1/Th2 balance, favoring a Th1 cell response. Several reports suggest that the makeup of intestinal microbiota can be different in individuals with allergic disorders and in those who reside in industrialized countries where the prevalence of allergy is higher [45–47]. For example, children from an industrialized country like Sweden harbor less lactobacilli and bifidobacteria in their bowels in comparison with children who live in countries like Estonia, where allergic disorders are not so common [48, 49].

The concept that children with allergic disorder harbor a different profile of microbiota has been supported by several other studies [14, 46, 50]. Not only can the composition of the intestinal microbiota vary but also the metabolic activity of the microbiota can be different. For example, Swedish children who are at a high risk of developing allergy were found to have significantly higher levels of fecal butyrate, isovalerate, and caproate than Estonian children, who have a low risk for developing allergies [51]. The KOALA study revealed that *Clostridium difficile* colonization at the age of 1 month was associated with an increased likelihood of eczema, recurrent wheezing, and atopic dermatitis. While this concept has been validated in several other studies, there are a few reports that do not show a significant difference in microbiota composition [52].

6.4 Probiotics and Prebiotics

Probiotics are defined as "live microorganisms" that, when administered in adequate amounts, confer a health benefit to the host. Examples of probiotics are *Lactobacillus rhamnosus* GG, *L. casei* Shirota, *L. johnsonii* La1, and *Bifidobacterium lactis* Bb12. Various health effects have been reported for probiotics, including some immune modulation activities [53]. Selected lactobacilli and bifidobacteria have been shown to be able to enhance the production of IgA [54, 55], while reduction in the production of IgE has been observed in mice [56].

Several animal and human studies have provided unequivocal evidence that specific strains of probiotics are able to stimulate, as well as regulate several aspects of natural and acquired immune responses [57]. Several probiotic effector molecules involved in the immune interactions have been identified, including bacterial cell wall components such as peptidoglycans and lipoteichoic acid, as well as specific proteins (reviewed in [23, 33]).

Allergy disorders are associated with a shift of the Th1/Th2 cytokine balance towards a Th2 response, including release of Th2 cytokines such as IL-4, IL-5, and IL-13, as well as IgE productions [58]. Probiotics can modulate the TLR and the proteoglycan recognition proteins of enterocytes, leading to the activation of DCs and a Th1 response. The resulting stimulation of Th1 cytokines can suppress Th2 responses [58]. Close contact of the probiotics with the intestinal mucosa may lead to an enhanced interaction of the probiotics and the intestinal immune system. This interaction may stimulate naive T cells to differentiate into Th1 cells under the influence of IFN-γ, IL-2, and IL-12 instead of Th2 cells. The result of this shift in T-cell differentiation from Th2 to Th1 is a reduced production of IgE and an increased secretion of IgA [59] which leads to a reduced allergic response.

Prebiotics are defined as nondigestible food ingredients, such as fructo-oligosaccharides and trans-beta-galacto-oligosaccharides and lactulose, that beneficially affect the host by selectively stimulating the growth and/or activity of one or a limited number of bacterial species (i.e., bifidobacteria and lactobacilli) already resident in the colon [60], thus producing a prebiotic effect.

The health effects of prebiotics are less well established than those for probiotics, and their effect on the immune system is still unknown [61]. However, because prebiotics influence the composition and activity of the normal microbiota and the microbiota is known to have a major effect on the immune system, it can be anticipated that prebiotics indirectly modulate the immune system [14]. The term "synbiotics" refers to the use of both prebiotics and probiotics simultaneously.

6.4.1   Probiotics in Prevention of Allergic Disease

In addition to treatment of allergy, it has been observed that selected probiotics can reduce the risk for the development of allergy. One of the earliest studies was performed with a nonpathogenic *E. coli* administered to term and preterm infants. At 10 and 20 years of age, children treated with *E. coli* suffered significantly fewer allergic diseases than the subjects in the control group [62]. In a recent study, the efficacy of *L. rhamnosus* GG on at-risk infants was studied; children of allergic mothers have about 50 % risk of developing allergy. Pregnant allergic mothers were given *L. rhamnosus* GG or a placebo from 2 to 4 weeks before the calculated date of delivery in a randomized double-blind trial. After delivery, the children received *L. rhamnosus* GG for 6 months. After 4 years, 46 % of the children in the placebo group had developed atopic eczema, whereas in the probiotics group this number was lowered to 26 % [63]. However, the serum IgE levels did not differ between the two groups. Few studies have addressed human mucosal responses to probiotics at the molecular level, and this represents an important observation in order to explain the probiotic effects observed at the clinical level. However, recent nutrigenomic approaches in healthy and diseased human volunteers illustrate the potential of post-genomic methodologies to decipher human mucosal responses to probiotics in relation to the potential clinical impact of these treatments, and such approaches have, in some cases, suggested novel, unexplored clinical effects of probiotics in humans [64]. For example, consumption of *L. rhamnosus* str. GG has been associated with prevention or relief of allergic symptoms [65, 66], possibly by preventing excess production of Th2 cells and maintaining a homeostatic Treg/Th1/Th2 cell ratio [67, 68]. Transcriptome studies of duodenal biopsies collected in a placebo-controlled, crossover-design trial that included *L. rhamnosus* str. GG consumption by healthy human volunteers demonstrates a remarkable correspondence between transcriptional modulation and the probiotic effects.

Despite a number of basic research data, probiotic clinical research in food allergy is still very modest, but the most recent evidence supports the potential clinical impact derived from a manipulation of intestinal microbiota in changing the pattern in cow's milk allergy, the most common food allergy in childhood [69].

To improve the application of probiotics to support and stimulate human health, it is important to improve our understanding of their mode of action. The field of probiotic research has progressed towards the molecular science of host–microorganism interactions, but will still require the comprehensive and detailed characterization of probiotic (immunomodulatory) molecules in relation to the molecular and physiological host responses that they can elicit. This will also clarify the species and strain specificity of probiotics, which has been poorly addressed to date, as only a few studies have compared multiple probiotic species and/or strains in a single experimental setup in vivo [23].

References

1. Bach JF. **The effect of infections on susceptibility to autoimmune and allergic diseases**. *N Engl J Med*. 2002, **347** (12):911–920.
2. Rook GAW. **99th Dahlem Conference on Infection, Inflammation and Chronic Inflammatory Disorders: Darwinian medicine and the 'hygiene' or 'old friends' hypothesis**. *Clin Exp Immunol*. 2010, **160** (1):70–79.
3. Graham-Rowe D. **Lifestyle: When allergies go west**. *Nature*. 2011, **479** (7374).
4. Wills-Karp M, Santeliz J, Karp CL. **The germless theory of allergic disease: revisiting the hygiene hypothesis**. *Nat Rev Immunol*. 2001, **1** (1):69–75.
5. Strachan DP. **Hay fever, hygiene, and household size**. *BMJ*. 1989, **299** (6710):1259–1260.
6. von Mutius E, Vercelli D. **Farm living: effects on childhood asthma and allergy**. *Nat Rev Immunol*. 2010, **10** (12):861–868.
7. Rook GAW. **Review series on helminths, immune modulation and the hygiene hypothesis: The broader implications of the hygiene hypothesis**. *Immunology*. 2009, **126** (1):3–11.
8. von Hertzen L, Hanski I, Haahtela T. **Natural immunity. Biodiversity loss and inflammatory diseases aretwo globalmegatrends that might be related**. *EMBO Rep*. 2011, **12** (11):1089–1093.
9. Hanski I, von Hertzen L, Fyhrquist N, Koskinen K, Torppa K, Laatikainen T, Karisola P, Auvinen P, Paulin L, Makela MJ, Vartiainen E, Kosunen TU, Alenius H, Haahtela T. **Environmental biodiversity, humanmicrobiota, and allergy are interrelated**. *Proc Natl Acad Sci U S A*. 2012, **109** (21):8334–8339.

10. Ouwehand AC. **Antiallergic Effects of Probiotics.** *J Nutr*. 2007, **137** (3):794S–797S.
11. Ouwehand AC, Derrien M, de Vos W, Tiihonen K, Rautonen N. Prebiotics and othermicrobial substrates for gut functionality. Curr Opin Biotechnol. 2005, 16 (2):212–217.
12. van Baarlen P, Troost F, van der Meer C, Hooiveld G, Boekschoten M, Brummer RJM, Kleerebezem M. **Human mucosal in vivo transcriptome responses tothree lactobacilli indicate how probiotics may modulate human cellular pathways.** *Proc Natl Acad Sci U S A*. 2010.
13. Brandtzaeg LH, Per. **Development and Function of Intestinal B and T Cells.** *Microb Ecol in Health D*. 2000, **12** (2):110–127.
14. Ouwehand A, Isolauri E, Salminen S. **The role of the intestinalmicroflora for the development of the immune system in early childhood.** *Eur J Nutr*. 2002, **41** Suppl 1:I32–37.
15. Tannock G. **Normalmicroflora: an introduction tomicrobes inhabiting the human body.** Chapman & Hall, London, 1995, 14–21.
16. Sanford PA. **Digestive System Physiology.** Edward Arnold, London, 1992, 46–72.
17. Lee A, Fox J, Hazell S. **Pathogenicity of Helicobacter pylori: a perspective.** *Infect Immun*. 1993, **61** (5):1601–1610.
18. Justesen T, Nielsen OH, Hjelt K, Krasilnikoff PA. **Normal cultivablemicroflora in upper jejunal fluid in children without gastrointestinal disorders.** *J Pediatr Gastroenterol Nutr*. 1984, 3 (5):683–686.
19. Gronlund MM, Lehtonen OP, Eerola E, Kero P. **Fecalmicroflora in healthy infants born by different methods of delivery: permanent changes in intestinal flora after cesarean delivery.** *J Pediatr Gastroenterol Nutr*. 1999, **28** (1):19–25.
20. Kleessen EB, Jaana Mättö, Brigitta. **Culture-Based Knowledge on Biodiversity, Development and Stability of Human GastrointestinalMicroflora.** *Microb Ecol in Health D*. 2000, **12** (2):53–63.
21. Patricia M, Heavey IRR. **The GutMicroflora of the Developing Infant:Microbiology and Metabolism.** *Microb Ecol in Health D*. 1999, **11** (2):75–83.
22. Benno Y, Endo K, Mizutani T, Namba Y, Komori T, Mitsuoka T. **Comparison of fecalmicroflora of elderly persons in rural and urban areas of Japan.** *Appl Environ Microbiol*. 1989, **55** (5):1100–1105.
23. Bron PA, van Baarlen P, Kleerebezem M. **Emerging molecular insights into the interaction between probiotics and the host intestinal mucosa.** *Nat Rev Microb*. 2011, **10** (1):66–78.
24. O'Hara AM, O'Regan P, Fanning A, O'Mahony C, Macsharry J, Lyons A, Bienenstock J, O'Mahony L, Shanahan F. **Functional modulation of human intestinal epithelial cell responses by Bifidobacterium infantis and Lactobacillus salivarius.** *Immunology*. 2006, **118** (2):202–215.
25. McCracken VJ, Lorenz RG. **The gastrointestinal ecosystem: a precarious alliance among epithelium, immunity andmicrobiota.** *Cell Microbiol*. 2001, 3 (1):1–11.
26. Liew FY. **T(H)1 and T(H)2 cells: a historical perspective.** *Nat Rev Immunol*. 2002, **2** (1):55–60.
27. Barnes MJ, Powrie F. **Regulatory T cells reinforce intestinal homeostasis.** *Immunity*. 2009, **31** (3):401–411.
28. Kelsall B. **Recent progress in understanding the phenotype and function of intestinal dendritic cells and macrophages.** *Mucosal Immunol*. 2008, **1** (6):460–469.
29. Rescigno M. **Intestinal dendritic cells.** *Adv Immunol*. 2010, **107**:109–138.
30. Coombes JL, Powrie F. **Dendritic cells in intestinal immune regulation.** *Nat Rev Immunol*. 2008, 8 (6):435–446.
31. Kapsenberg ML. **Dendritic-cell control of pathogen-driven T-cell polarization.** *Nat Rev Immunol*. 2003, 3 (12):984–993.
32. Sansonetti PJ, Medzhitov R. **Learning Tolerance while Fighting Ignorance.** *Cell*. 2009, **138** (3):416–420.
33. Kleerebezem M, Hols P, Bernard E, Rolain T, Zhou M, Siezen RJ, Bron PA. **The extracellular biology of the lactobacilli.** *FEMS Microbiol Rev*. 2010, **34** (2):199–230.
34. Lebeer S, Vanderleyden J, De Keersmaecker SC. **Host interactions of probiotic bacterial surface molecules: comparison with commensals and pathogens.** *Nat Rev Microbiol*. 2010, 8 (3):171–184.

35. Artis D. **Epithelial-cell recognition of commensal bacteria and maintenance of immune homeostasis in the gut.** *Nat Rev Immunol.* 2008, **8** (6):411–420.
36. Reid G, Younes JA, Van der Mei HC, Gloor GB, Knight R, Busscher HJ. **Microbiota restoration: natural and supplemented recovery of humanmicrobial communities.** *Nat Rev Microbiol.* 2011, **9** (1):27–38.
37. Spiekermann GM, Walker WA. **Oral tolerance and its role in clinical disease.** *J Pediatr Gastroenterol Nutr.* 2001, **32** (3):237–255.
38. Hanson LÅ, Dahlman-Höglund A, Karlsson M, Lundin S, Dahlgren U, Telemo E. **Normalmicrobial flora of the gut and the immune system.** Hanson LÅ, Yolken RH (eds) Nestlé Workshop Series, Vol 42. Vevey/Lippincott- Raven Publishers, Philadelphia, 1999, 217–228.
39. Holgate ST. **The epidemic of allergy and asthma.** *Nature.* 1999, **402** (6760 Suppl):B2–4.
40. von der Weid T, Bulliard C, Schiffrin EJ. **Induction by a Lactic Acid Bacterium of a Population of CD4 + T Cells with Low Proliferative Capacity That Produce Transforming Growth Factor β and Interleukin-10.** *Clin Diagn Lab Immun.* 2001, **8** (4):695–701.
41. Kuitunen M. **Probiotics and prebiotics in preventing food allergy and eczema.** *Curr Opin Allergy Clin Immunol.* 2013, **13** (3):280–286.
42. Michail S. **The role of probiotics in allergic diseases.** *Allergy Asthma Clin Immunol.* 2009, **5** (1):5.
43. Flohr C, Pascoe D, Williams HC. **Atopic dermatitis and the 'hygiene hypothesis': too clean to be true?** *Brit J Dermatol.* 2005, **152** (2):202–216.
44. Ogden NS, Bielory L. **Probiotics: a complementary approach in the treatment and prevention of pediatric atopic disease.** *Curr Opin Allergy Clin Immunol.* 2005, **5** (2):179–184.
45. Kirjavainen PV, Arvola T, Salminen SJ, Isolauri E. **Aberrant composition of gutmicrobiota of allergic infants: a target of bifidobacterial therapy at weaning?** *Gut.* 2002, **51** (1):51–55.
46. Kirjavainen PV, Apostolou E, Arvola T, Salminen SJ, Gibson GR, Isolauri E. **Characterizing the composition of intestinalmicroflora as a prospective treatment target in infant allergic disease.** *FEMS Immunol Med Mic.* 2001, **32** (1):1–7.
47. Björkstén B. **The gastrointestinal flora and the skin—is there a link?** *Pediatr Allergy Immu.* 2001, **12**:51–55.
48. Sepp E, Julge K, Mikelsaar M, Bjorksten B. **Intestinalmicrobiota and immunoglobulin E responses in 5-year-old Estonian children.** *Clin Exp Allergy.* 2005, **35** (9):1141–1146.
49. Bjorksten B, Sepp E, Julge K, Voor T, Mikelsaar M. **Allergy development and the intestinalmicroflora during the firstyear of life.** *J Allergy Clin Immunol.* 2001, **108** (4):516–520.
50. Kalliomaki M, Isolauri E. **Role of intestinal flora in the development of allergy.** *Curr Opin Allergy Clin Immunol.* 2003, **3** (1):15–20.
51. Norin E, Midtvedt T, Björksten B. **Development of faecal short-chain fatty acid pattern during the firstyear of life in estonian and swedish infants.** *Microb Ecol in Health D.* 2004, **16** (1):8–12.
52. Adlerberth I, Strachan DP, Matricardi PM, Ahrne S, Orfei L, Aberg N, Perkin MR, Tripodi S, Hesselmar B, Saalman R, Coates AR, Bonanno CL, Panetta V, Wold AE. **Gut microbiota and development of atopic eczema in 3 European birth cohorts.** *J Allergy Clin Immunol.* 2007, **120** (2):343–350.
53. Salminen S, Ouwehand A, Benno Y, Lee YK. **Probiotics: how should they be defined?** *Trends Food Sci Tech.* 1999, **10** (3):107–110.
54. Majamaa H, Isolauri E, Saxelin M, Vesikari T. **Lactic acid bacteria in the treatment of acute rotavirus gastroenteritis.** *J Pediatr Gastroenterol Nutr.* 1995, **20** (3):333–338.
55. Yasui H, Nagaoka N, Mike A, Hayakawa K, Ohwaki M. **Detection of Bifidobacterium Strains that Induce Large Quantities of IgA.** *Microb Ecol in Health D.* 1992, **5** (3):155–162.
56. Matsuzaki T, Yamazaki R, Hashimoto S, Yokokura T. **The effect of oral feeding of Lactobacillus casei strain Shirota on immunoglobulin E production in mice.** *J Dairy Sci.* 1998, **81** (1):48–53.
57. Gill H, Prasad J. **Probiotics, immunomodulation, and health benefits.** *Adv Exp Med Biol.* 2008, **606**:423–454.

58. Winkler P, Ghadimi D, Schrezenmeir J, Kraehenbuhl J-P. **Molecular and Cellular Basis of Microflora-Host Interactions.** *J Nutr.* 2007, **137** (3):756S–772S.
59. Tejada-Simon MV, Pestka JJ. **Proinflammatory cytokine and nitric oxide induction in murine macrophages by cell wall and cytoplasmic extracts of lactic acid bacteria.** *J Food Prot.* 1999, **62** (12):1435–1444.
60. Gibson GR, Roberfroid MB. **Dietary modulation of the human colonic microbiota: introducing the concept of prebiotics.** *J Nutr.* 1995, **125** (6):1401–1412.
61. Roberfroid MB. **Prebiotics and probiotics: are they functional foods?** *Am J Clin Nutr.* 2000, **71** (6):1682s–1687s.
62. Lodinova-Zadnikova R, Cukrowska B, Tlaskalova-Hogenova H. **Oral administration of probiotic Escherichia coli after birth reduces frequency of allergies and repeated infections later in life (after 10 and 20 years).** *Int Arch Allergy Immunol.* 2003, **131** (3):209–211.
63. Kalliomaki M, Salminen S, Poussa T, Arvilommi H, Isolauri E. **Probiotics and prevention of atopic disease: 4-year follow-up of a randomised placebo-controlled trial.** *Lancet.* 2003, **361** (9372):1869–1871.
64. van Baarlena P, Troosta F, van der Meera C, Hooivelda G, Boekschotena M, . Brummera RJM, Kleerebezema M. **Human mucosal in vivo transcriptome responses to three lactobacilli indicate how probiotics may modulate human cellular pathways.** *Proc. Natl Acad. Sci. USA.* 2011, **108** (1):4562–4569.
65. Kalliomäki M, Salminen S, Arvilommi H, Kero P, Koskinen P, Isolauri E. **Probiotics in primary prevention of atopic disease: a randomised placebo-controlled trial.** *Lancet* 2001, **357** (9262):1076–1079.
66. Kalliomaki M, Salminen S, Poussa T, Arvilommi H, Isolauri E. **Probiotics and prevention of atopic disease: 4-year follow-up of a randomised placebocontrolled trial.** *Lancet* 2003, **361** (9372):1869–1871.
67. Kalliomaki M, Isolauri E. **Role of intestinal flora in the development of allergy.** *Curr. Opin. Allergy Clin. Immunol.* 2003, **3** (1):15–20.
68. Schultz M, Linde HJ, Lehn N, Zimmermann K, Grossmann J, Falk W, Schölmerich J. **Immunomodulatory consequences of oral administration of Lactobacillus rhamnosus strain GG in healthy volunteers.** *J. Dairy Res.* 2003, **70** (2):165–173.
69. Canani R, Di Costanzo M. **Gut Microbiota as Potential Therapeutic Target for the Treatment of Cow's Milk Allergy.** *Nutrients.* 2013, **5** (3):651–662.

Chapter 7
Phytochemicals and Hypersensitivity Disorders

Contents

Abbreviations

$\Delta\Psi_m$	Mitochondrial membrane potential
EC	Epicatechin
EGCG	Epigallocatechin gallate
FcεRI	High-affinity IgE receptor
GIT	Gastrointestinal tract
IL-2R	IL-2 receptor
IL-4R	IL-4 receptor
JAK	Janus kinase
LPS	Lipopolysaccharide
MAPK	Mitogen-activated protein kinase
MDC	Monocyte-derived chemokine
NF-κB	Nuclear transcription factor-kappa B
PUFA	Omega-3 polyunsaturated fatty acids
ROS	Reactive oxygen species
SOCE	Store-operated Ca^{2+} entry

T. Ćirković Veličković, M. Gavrović-Jankulović, *Food Allergens*,
Food Microbiology and Food Safety, DOI 10.1007/978-1-4939-0841-7_7,
© Springer Science+Business Media New York 2014

STAT   Signal transducer and activator of transcription
TJ   Tight junction
ZAP-70   ξ-associated 70-kDa protein

Summary It has been shown that phytochemicals may act on allergic disease either during allergic sensitization or on consolidated disease. There is a renewed interest in the search for new phytochemicals that could be developed as useful anti-inflammatory and antiallergic agents to reduce the risk of many diseases. A good number of plant products with anti-inflammatory and antiallergic activities have been documented, but very few of these compounds have reached clinical use and there is scant scientific evidence that could explain their mode of action. The activation of nuclear transcription factor-kappa B (NF-κB) has now been linked to a variety of inflammatory diseases, while data from numerous studies underline the importance of phytochemicals in inhibiting the pathway that activates this transcription factor.

Phytochemicals, especially phenolics, show both anti-inflammatory and antiallergic activities in vitro and in vivo. Several cellular action mechanisms are proposed to explain their mode of action. However, any single mechanism could not explain all of their in vivo activities. Possible mechanisms involve interference of polyphenols with antigen-presenting cell maturation, inhibition of Th2-type cytokine signaling and secretion, release of mediators of allergic inflammation, as well as direct effects of dietary polyphenols on food allergen solubility, digestion process, and intestinal barrier function.

7.1   Diet and Hypersensitivity Disorders

The interaction between genetic and environmental factors is generally accepted to cause individuals to be sensitized with environmental allergens and to suffer from allergic diseases. Recent changes in the environment might have contributed to the increase in hypersensitivity disorders. It is of importance to reveal which environmental factors cause such high prevalence and to find strategies to prevent their development. The change of diet is considered to be one of the environmental factors that might be responsible for such an increase [1].

Foods include both allergy-promoting and antiallergic nutrients. Vitamins A, C, E, selenium, and copper are antioxidants and vitamin C and E also have other anti-inflammatory and antiallergic effects. Omega-3 polyunsaturated fatty acids (PUFA) stabilize the mast cell membrane and decrease leukotriene (LT) C4 synthesis, whereas omega-6 PUFA are precursors for LT C4 and thus may promote allergic inflammation. Based on the activity of the nutrients and the epidemiological studies, dietary manipulation of these nutrients may ameliorate allergic symptoms. It has been reported that reduced consumption of foods containing antioxidants (fruits and vegetables), increased omega-6 PUFA intake, and reduced omega-3 PUFA intake have been implicated for the increase in asthma and atopic diseases. The authors also concluded that intake of dietary antioxidant and lipid during pregnancy and early childhood might decrease the onset of allergic diseases [1].

Administration of flavonoids into atopic dermatitis-prone mice showed a preventative and ameliorative effect. Recent epidemiological studies reported that a low incidence of asthma was significantly observed in a population with a high intake of flavonoids [2].

However, intervention studies so far have reached no consistent conclusion. It also seems likely that there is individual variation in the responses of individuals to lipid, and probably antioxidant supplementation [1].

7.2 Dietary Phytochemicals

Dietary phytochemicals can be classified into: carotenoids, phenolics, alkaloids, nitrogen-containing compounds, and organosulfur compounds [3]. Phenolics represent the largest and the most studied group of dietary phytochemicals (Fig. 7.1).

Many in vitro, in vivo, and epidemiological studies have suggested that dietary phytochemicals, especially polyphenols, have beneficial effects on human health, and treatment and prevention of cardiovascular disease and cancer [5–8]. Intake of polyphenol-rich beverages, especially green tea, have beneficial effects on many chronic and inflammatory diseases, including heart disease, diabetes, neurodegenerative disease, arthritis, and cancer [7, 9, 10]. These beneficial effects are mainly due to the antioxidant and anti-inflammatory properties of polyphenols [6, 11]. The inhibition of digestive enzymes involved in carbohydrate, lipid, and protein metabolism by dietary polyphenols may be another important mechanism for the health benefits attributed to a diet rich in fruit and vegetables [5, 12, 13]. The focus for many nutritionists is also the research of food compounds that may influence the functionality of the immune system.

The available epidemiological, animal, and molecular data suggest that there are associations between antioxidants and asthma and, to a much lesser extent, atopic dermatitis and atopic rhinitis. However, the exact nature of the relationships and the potential for therapeutic intervention remain unclear [14].

A clinical open study was performed on the benefits of a typical vegetarian diet on the onset of allergic symptoms in adult patients with atopic dermatitis. The diet consisted of fruits and vegetables. After a 2-month period of treatment, the severity of dermatitis decreased from 49.9 ± 18.6 to 27.4 ± 16.8 based on a score of atopic dermatitis severity, the SCORAD index, and on serological parameters including lactate dehydrogenase-5 activity and a number of peripheral eosinophils [15]. One of the characteristics of this diet was a high intake of *flavonoids*. By this vegetarian diet, it was calculated that 17 mg of apigenin, 1.6 mg of luteolin, 19.5 mg of quercetin, and 29 mg of kaempferol were consumed daily.

In particular, phytochemicals such as polyphenols that exhibit a strong immunomodulatory activity may influence initiation and maintenance of allergic inflammation by influencing:

- Antigen presentation and maturation of antigen-presenting cells
- Release of mediators of allergic inflammation from effector cells
- Immune system regulatory mechanisms, IgE and cytokine production

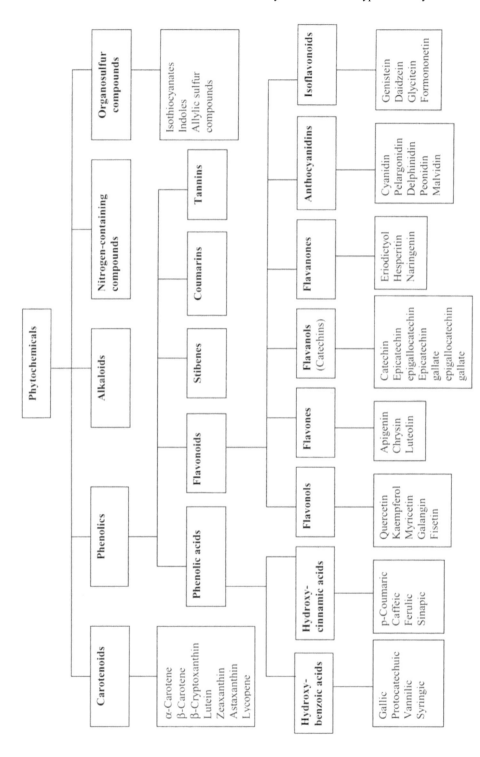

Fig. 7.1 Classification of dietary phytochemicals [4]

Additionally, interactions of food proteins and polyphenols may influence:

* Food allergen digestibility and solubility
* Allergen bioavailability and uptake in the gastrointestinal tract (GIT)

Moreover, protein conformation can be modified by binding to polyphenols and thus influence allergenic protein–IgE interactions, in case conformational IgE-binding epitopes are modified [16, 17].

7.2.1 Flavonoids

Flavonoids represent the most studied class of polyphenols. Flavonoids comprise a large group of low molecular weight polyphenolic secondary plant metabolites and are found in vegetables, fruits, nuts, seeds, stems, flowers, roots, bark, tea, wine, and coffee and are thus common substances in our daily diet. It is estimated that flavonoids account for approximately two thirds of the phenolics in our diet and the remaining one third are from phenolic acids [4]. Flavonoids have been recognized to exert antioxidant, antibacterial, and antiviral activity, and possess anti-inflammatory, analgesic, hepatoprotective, cytostatic, apoptotic, estrogenic or antiestrogenic properties, as well as antiallergic effects.

Based on their skeleton, flavonoids are categorized into eight groups: *flavans, flavanones, isoflavanones, flavones, isoflavones, anthocyanidins, chalcones, and flavonolignans.*

Their skeleton is a heterocyclic hydrocarbon, chromane (Fig. 7.2), and substitution of its ring C in position 2 or 3 with a phenyl group (ring B) results in flavans or isoflavans. An oxo group in position 4 leads to flavanones and isoflavanones. The presence of a double bond between C2 and C3 provides flavones and isoflavones. An additional double bond between C1 and C2 makes these compounds colorful anthocyanidins.

Natural flavonoids usually occur as glycosides (e.g., glucosides, rhamnoglucosides, and rutinosides).

Flavonols such as quercetin, kaempferol, galangin, morin, rutin, myricetin, isorhamnetin, and isoquercetin can be found in onions, apples, berries, kale, leeks, broccoli, blueberries, red wine, and tea.

Flavones such as luteolin, chrysin, and apigenin are commonly found in fruit skins, parsley, and celery.

Isoflavones such as genistein, daidzein, and glycitein are present in leguminous plants, mainly soy and soy products.

Flavanones such as naringenin and hesperidin are exclusive to citrus fruits.

Flavanols include monomers such as epicatechin (EC), catechin, gallocatechin, epigallocatechin, and epigallocatechin gallate (EGCG), and also their polymers called proanthocyanidins.

Proanthocyanidins or condensed tannins consist of procyanidins (polymers of EC and catechin, found in cocoa), prodelphinidins (polymers of epigallocatechin or gallocatechin), and propelargonidins (epiafzelechin or afzelechin polymers).

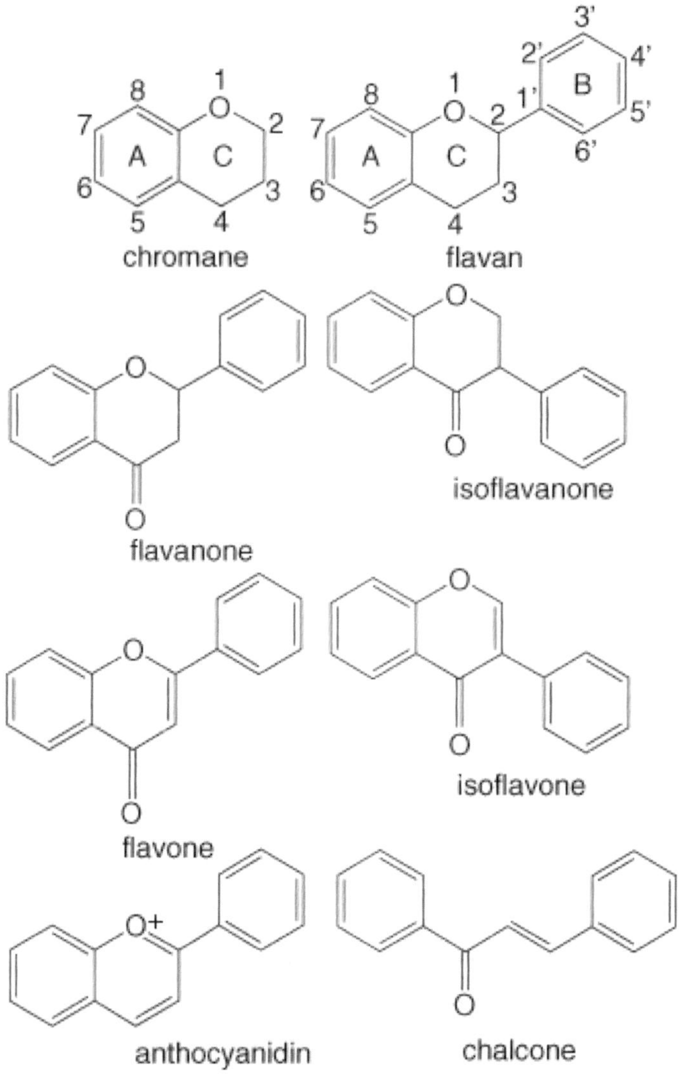

Fig. 7.2 Basic structure of flavonoids

Anthocyanidins include pelargonidin, cyanidin, and malvidin present in berry fruits and red wine.

Most studies evaluating the effects of flavonoids on the immune system are performed in vitro. These studies allow one to approach the molecular mechanisms and cellular targets of flavonoids. Few studies evaluated immunological effects of flavonoids in vivo in order to better reflect the effects of these compounds after absorption and metabolism. The most efficient and best studies on groups of flavonoids in hypersensitivity disorders are flavanols (catechins), flavonols, and flavones.

7.2.1.1   Antiallergic Activities of Flavanols (Catechins)

Flavanols, especially catechins of green tea, have been thoroughly studied for their beneficial effects in immune system disorders.

Tea catechins are characterized by the dihydroxyl or trihydroxyl substitutions on the B ring and the m-5,7-dihydroxyl substitutions on the A ring (Fig. 7.2). The B ring seems to be the principal site of antioxidant reactions and the antioxidant activity is further increased by the trihydroxyl structure in the gallate ring (gallate) in EGCG. The polyphenolic structure allows electron delocalization, conferring the ability to quench free radicals. Tea preparations have been shown to react with reactive oxygen species (ROS) such as superoxide radical, singlet oxygen, hydroxyl radical, peroxyl radical, nitric oxide, nitrogen dioxide, and peroxynitrite.

Among tea catechins, EGCG is the most effective in reacting with the majority of ROS. Tea polyphenols are also strong chelators of metal ions; the chelation of free metal ions prevents the formation of ROS from the auto-oxidation of many compounds. The vicinal dihydroxy or trihydroxy structures not only contribute to the antioxidative activity of tea polyphenols but also increase the susceptibility of these compounds to air oxidation under alkaline or neutral pH. In the case of EGCG, auto-oxidation generates superoxide anion and hydrogen peroxide and leads to the formation of dimers, such as theasinensins. These reactions that occur under cell culture conditions are due to superoxide anion-catalyzed chain reactions.

The polyphenolic structure of tea polyphenols also makes them good donors for hydrogen bonding. For example, hydrogen bonding of water molecules to EGCG forms a large hydration shell, which reduces the absorption of EGCG. This hydrogen-bonding capacity also enables tea polyphenols to bind strongly to proteins and nucleic acids. How these effects contribute to the health-promoting effects of green tea catechins is partly determined by the bioavailability of these compounds *in vivo*.

Previously, EGCG has been shown to bind to salivary proline-rich proteins, fibronectin, fibrinogen- and histidine-rich glycoproteins, 67-kDa laminin receptor, and Bcl-2 proteins. Using an EGCG-Sepharose 4B column, two-dimensional electrophoresis, and matrix-assisted laser desorption/ionization-time-of-flight mass spectroscopy, Dong *et al.* identified vimentin, insulin-like growth factor 1 receptor, 78-kDa glucose-regulated protein, and ζ-associated 70-kDa protein (ZAP-70) as high-affinity EGCG-binding proteins.

The ZAP-70 is a tyrosine kinase of the spleen tyrosine kinase (Syk) family, which is associated with the ζ subunit of the T-cell receptor (TCR). A direct binding of EGCG into the ATP-binding pocket of ZAP-70 led to the inhibition of the kinase activity and downstream events associated with kinase inhibition [18]. EGCG also binds to several membrane proteins, such as CD11b [19, 20], CD4 [21], 67-kDa laminin receptor [22], and protein kinases [23].

Although the exact beneficial dosages of green tea polyphenols are still a matter of dispute, it is generally believed that green tea catechins and its major component EGCG, in particular, have strong anti-inflammatory and immunosuppressive effects in vitro and in vivo. One mechanism of the anti-inflammatory effect of EGCG is the attenuation of the adhesive and migratory properties of CD8$^+$ T cells by the direct binding of EGCG to CD11b [20]. EGCG also induced apoptosis of human monocytes and dendritic cells (DCs), another possible mechanism of its anti-inflammatory and immunosuppressive effects [24, 25].

Although EGCG is a principal green tea component, due to limited bioavailability the physiological concentration of EGCG in human plasma following regular tea intake is reported to be below 1 μmol/L. Higher plasma levels of EGCG can be reached following consumption of pharmacological doses of green tea supplements. Oral administration of 525 mg of EGCG in healthy volunteers was shown to give the maximum human plasma level of 4.4 μmol/L (2 μg/mL) of EGCG [26]. A single oral dose of 1,600 mg of EGCG under fasting conditions gave maximum plasma levels of 3.4 μg/mL of EGCG. [27]

Many in vitro studies have been performed with EGCG concentrations that were much above physiological level. New evidences have shown that a wide range of immune cells was targeted by green tea polyphenols and its major active component EGCG, in particular. The described effects include impact on differentiation and maturation of DCs [25, 28], inhibition of histamine release and cytokine secretion by basophils [29, 30], prevention of mast cell degranulation [31], effects on the functionality and number of T regulatory cells [28, 32], human monocyte vitality, adhesion, and mobility [24, 33], and IL-17 and TNF-alpha expression in human Th17 cells [34].

Very potent immune modulating activities of EGCG have been associated with its interference with CD25 (Interleukin-2 receptor, IL-2R), that prevents IL-2 binding [35]. IL-2 is a key cytokine involved in T-cell proliferation and activation and controls secretion of many other cytokines, such as IL-10 and IL-5. The immunosuppressive effect of EGCG is also shown for CD4$^+$, CD8$^+$ cell subsets in mice [36], a polyclonal stimulation of human T cells [37], a reduced utilization of IL-2, and cell cycle arrest in both CD4$^+$ and CD8$^+$ cell subsets in mice [36].

The beneficial antiallergic activities of green tea intake have also been described in vivo, in respiratory allergic patients [2], and in animal models of house dust mites-induced atopic dermatitis [38]. The main antiallergic properties of the green tea catechins are associated with the effect on effector cells, basophils, and their tissue counterparts, mast cells, via direct inhibition of allergic degranulation, and secretion of inflammatory mediators [29, 31].

A recent study demonstrated downregulation of gamma common chain receptor subunit (γc) of several regulatory cytokine receptors (IL-2, IL-7, and IL-15) on CD4$^+$ T cells by EGCG, thus impairing signalization of several members of γc/JAK3-dependent cytokines [39]. Impairment of IL-2/IL-2R signaling by EGCG has been demonstrated by a reduced phosphorylation of signal transducer and activator of transcription 5 (STAT5) in the presence of EGCG due to the effects on the γc expression.

On the contrary, potential pro-allergic and pro-inflammatory activities of high doses of EGCG have been described in Jurkat T cell lines, through upregulation of Th2 cytokines, especially IL-5 [40]. Recent evidence has shown that EGCG also promoted inflammatory response [41] in mice receiving a high bolus dose of EGCG.

7.2.1.2   Antiallergic Activities of Flavonols

Kaempferol, which belongs to the flavonol group, is a strong antioxidant among natural flavonoids and is the essential component of many beverages and vegetables. Kaempferol also showed a potent activity in protecting and preventing IgE-mediated hypersensitivity disorders [42–44]. It has been shown that dietary kaempferol is effective in ameliorating allergic and inflammatory airway diseases through disturbing nuclear transcription factor-kappa B (NF-κB) signaling. Oral administration of kaempferol attenuated ovalbumin (OVA) challenge-elevated expression of eotaxin-1 and eosinophil major basic protein via the blockade of NF-κB transactivation, thereby blunting eosinophil accumulation in airway and lung tissue in mice with allergic asthma [44].

IL-4 is a crucial cytokine of IgE-mediated hypersensitivity disorders as it is essential for class switching in B cells and production of IgE. IL-4 signaling through a type I IL-4R involves receptor dimerization and transphosphorylation of Janus kinase (JAK)1 and JAK3 on signature tyrosine residues. IL-4 activates multiple signal transduction pathways in immune cells, with STAT6 being the crucial transduction molecule for biological functions of IL-4. Some pharmaceuticals target STAT6 activity for allergic disease prevention and treatment [45]. The phosphotyrosine motif on IL-4Rα acts as a docking site for STAT6 leading to the phosphorylation on tyrosines of STAT6 by JAKs [46]. Subsequently, phosphorylated STAT6 departs from the receptor, dimerizes, and translocates into the nucleus, where it turns on the expression of IL-4 target genes [46].

It has been demonstrated that kaempferol exhibits an inhibitory effect on STAT6 activation that correlates well with the inhibition of cell responses to IL-4 cytokine [47]. γc and JAK3 have previously been described as possible target molecules of antiallergic actions of *apigenin and naringenin* [48].

Recent data demonstrated that kaempferol significantly inhibited the lipopolysaccharide (LPS)-induced production of monocyte-derived chemokine (MDC), interferon gamma-induced protein 10 (IP-10), and IL-8 in human monocyte THP-1 cells [42]. Kaempferol was also able to suppress LPS-induced mitogen-activated protein kinase (MAPK) pathways.

Both *quercetin and kaempferol* effectively suppressed the development of IgE-mediated allergic inflammation of intestinal cell models [43]. Both flavonols inhibited the secretion of allergic mediators in RBL-2H3 cells and suppressed the CD23 mRNA expression and p38 MAPK activation in IL-4-stimulated Caco-2 cells. Flavonols also suppressed IgE-OVA-induced extra signal-regulated protein kinase activation and chemokine release.

Flavonols also affect IL-4, IL-13, and CD40 ligand expression by basophils. It has been demonstrated that the flavonol *fisetin* suppresses the expression of Th2-type cytokines (IL-4, IL-13, and IL-5) by basophils [49].

7.2.1.3   Antiallergic Activities of Flavones

The flavones luteolin and apigenin in a dose-dependent manner suppressed CD40 ligand expression by basophils. It has been shown that these inhibitory activities of luteolin on the expression of IL-4, IL-13, and CD40 ligand were accompanied with its suppressive activity of mRNA expression of IL-4, IL-13, and CD40 ligand [50].

Flavones (apigenin and chrysin) inhibited IL-4-induced ε germ line transcription which is essential for IgE class switching, and the phosphorylation of STAT6, JAK3, and IL-4Rα, whereas IL-4 signaling mediated through type II IL-4R was unaffected by flavones [48]. Flavones significantly reduce the cell surface and protein expression of γc, suggesting that γc may be a target molecule for the inhibitory effect of flavones on IL-4 signaling.

7.3 Mechanisms of Polyphenol Interference with Hypersensitivity Disorders

7.3.1 Effects of Polyphenols on Antigen Presentation and Maturation of Antigen-Presenting Cells

Studies have shown that both quercetin and EGCG were able to inhibit the expression of MHC class II and co-stimulatory molecules, such as CD11c, CD40, CD80, CD83, and CD86, in activated DCs [25, 51].

EGCG also induces apoptosis of human monocytes and DCs which may contribute to the anti-inflammatory and immunosuppressive effects of this flavanol [24, 25]. The described effects also include its impact on differentiation and maturation of DCs [25], human monocyte vitality, adhesion, and mobility [24, 33].

Ingestion of *resveratrol,* a polyphenol of red grapes structurally related to flavonoids, prevented the development of a food allergy model in mice. Given the in vitro findings, resveratrol might do so by inhibiting DC maturation and subsequent early T-cell activation and differentiation via downregulation of cholera toxin-induced cyclic adenosine monophosphate (cAMP) activation in mice. These results suggest that resveratrol may have potential for prophylaxis against food allergy [52].

7.3.2 Effects of Polyphenols on Release of Mediators of Allergic Inflammation

Mast cells and basophils express high-affinity IgE receptor (FcεRI) and play an important role in allergic inflammation through release of chemical mediators such as histamine, cysteinyl LTs, cytokines, and chemokines upon allergen cross-linking of FcεRI.

It has been shown that flavonoids inhibit histamine release, synthesis of IL-4 and IL-13, and CD40 ligand expression by basophils. Analyses of structure–activity relationships of 45 flavonoids and related compounds showed that luteolin, ayanin, apigenin, and fisetin were the strongest inhibitors of IL-4 production with an IC_{50} value of 2–5 μM and determined a fundamental structure for the inhibitory activity. The inhibitory activity of flavonoids on IL-4 and CD40 ligand expression was possibly mediated through their inhibitory action on activation of nuclear factors of activated T cells [2].

Flavonoids are inhibitors of chemical mediator release and cytokine production by mast cells and basophils, although their antiallergic action can also be due their effect on signal transducers in cells and dysregulation of cellular signaling by several cytokines, including IL-2 and regulatory cytokines, such as IL-7 and IL-15 [39]. The antiallergic effect of flavone inhibition of transport ATPase in histamine secretion from rat mast cells was demonstrated a long time ago [53]. Subsequently, it has been shown that quercetin, a naturally occurring flavonol structurally related to the antiallergic drug disodium cromoglycate was able to inhibit anaphylactic histamine release from rat intestinal mast cells [54] and antigen-stimulated human basophils [55].

The structure–activity relationship of flavonoids for antiallergic actions was studied by determining the IC_{50} values for the degranulation. The hexosaminidase release from RBL-2H3 cells was employed as an estimate for the antiallergic actions of flavonoids. Among 22 flavonoid compounds tested, luteolin, apigenin, diosmetin, fisetin, and quercetin were found to be most active with IC_{50} values less than 10 μM [56]. Flavonoids have also been shown to suppress cysteinyl LT synthesis through an inhibition of phospholipase (PL) A2 and/or 5-lipoxygenase (5LO) [2, 57].

7.3.3 Effects of Polyphenols on Immune System Regulatory Mechanisms, IgE and Cytokine Production

Allergic diseases are characterized by overproduction of IgE. IgE is produced by B cells. For the differentiation of B cells into IgE-producing cells, both the interaction of CD40 ligand with CD40 and the action of IL-4 or IL-13 on B cells are required.

The effect of flavonoids on IL-4, IL-13, and CD40 ligand expression was examined by basophils. The flavonoid fisetin suppresses the expression of Th2-type cytokines (IL-4, IL-13, and IL-5) by basophils [49]. Among the flavonoids examined, kaempferol and quercetin showed substantial inhibitory activities in cytokine expression but less so than those of fisetin.

It has been shown that luteolin, apigenin, and fisetin in a dose-dependent manner suppressed CD40 ligand expression by basophils, whereas myricetin even at 30 μM did not possess such activity. These inhibitory activities of luteolin on the expression of IL-4, IL-13, and CD40 ligand were accompanied with its suppressive activity of mRNA expression of IL-4, IL-13, and CD40 ligand [50]. Therefore, flavonoids such as *luteolin, apigenin, and fisetin are considered as potential natural IgE inhibitors*.

Flavones (*apigenin and chrysin*) inhibited *IL-4-induced ε germ line transcription* which is essential for IgE class switching and the phosphorylation of *STAT6, JAK3,* and *IL-4Rα,* whereas IL-4 signaling mediated through type II IL-4R was unaffected by flavones [48]. Flavones significantly reduce the cell surface and protein expression of γc, suggesting that γc is a target molecule for the inhibitory effect of flavones on IL-4 signaling. However, the potency of flavones to inhibit STAT6 and JAK3 phosphorylations is stronger than their potency to reduce γc expression.

It has also been reported that some natural compounds such as strictinin, apigenin, chrysin, polyphenol in tea, and astragalin, a major constituent of flavonoids in persimmon leaf extract, inhibit antigen-specific IgE production [48]. The potency of EGCG to inhibit IgE synthesis was also studied in a different experimental setup.

7.3.4 ROS Signaling and Polyphenols

Polyphenols can exert their antioxidative effects by scavenging ROS, or chelating transition metals. Polyphenols can also be potent pro-oxidants in vitro and in vivo, leading to the formation of ROS, such as hydrogen peroxide, the hydroperoxyl radical, and superoxide anion radical [7, 58].

ROS are involved as secondary messengers in many transduction pathways and regulate expression of other chemokines, such as Th1- and Th2-type cytokines [59–61]. The role which polyphenols play in the immune system has been initially attributed to their redox properties and an ability to scavenge ROS and transit metal ions.

ROS are involved as secondary messengers in T-cell receptor activation and other transduction pathways and regulate expression of Th1- and Th2-type cytokines [59]. It has been shown that major green tea catechin EGCG, through dismutation reaction in the presence of molecular oxygen, may increase ROS levels in mast cells and also augment antigen-induced ROS increase and mitochondrial membrane potential ($\Delta\Psi_m$) collapse [31]. It also inhibits antigen-induced Ca^{2+} influx and store-operated Ca^{2+} entry (SOCE), the principal mode of Ca^{2+} influx into mast cells. EGCG, but not EC, inhibited antigen-induced degranulation, LT C_4 secretion, and Ca^{2+} influx. EGCG also blocks SOCE without reducing Ca^{2+} store emptying, whereas EC does not, although it does reduce Ca^{2+} store emptying. EGCG also evokes intracellular ROS production, $\Delta\Psi_m$ collapse, cardiolipin oxidation, and mitochondrial Ca^{2+} ($[Ca^{2+}]_m$) release. Thus, it has been demonstrated that ROS production and $\Delta\Psi_m$ collapse are important mechanisms underlying the antiallergic effects of EGCG [31].

7.3.5 Interactions of Polyphenols with Dietary Proteins

Noncovalent interactions of polyphenols and globular proteins may result in complexation, stabilization of protein structure, protein unfolding, and precipitation [62]. It has been shown that the strength of interactions depends on the size of polyphenols, polyphenol structure, and amino acid sequence of proteins [63].

The structural characterization of the interactions between globular food proteins and polyphenols is a major step in elucidating the effect of polyphenols on globular protein structure. Tryptophan (Trp) residues are intrinsic fluorophores in proteins and spectroscopic techniques measuring quenching of Trp fluorescence in proteins have been applied to studies of polyphenol binding to various proteins such as milk caseins and β-lactoglobulin, lysozyme, hemoglobin, serum albumins, and gamma globulins. The quantum yield and emission maximum wavelength associated with intrinsic Trp fluorescence are very sensitive to the polarity of the environment and the structural changes in biomacromolecules.

Complexation of polyphenols and proteins can affect antioxidant activity of polyphenols by affecting their electron donation capacity and reducing the number of hydroxyl groups available in the solution. However, due to the prolonged life of polyphenols in the complex, the effect of complexation may be beneficial for the overall antioxidant activity of the polyphenols.

Few recent studies demonstrated new important biological effects of protein–polyphenol noncovalent interactions. Polyphenols exhibited a potent dose-dependent inhibitory activity on alpha-synuclein aggregation and were capable of disaggregating the preformed alpha-synuclein oligomers. A positive correlation was also found between the degree of polyphenol oligomerization and inhibition of elastase due to an increased number of protein interacting groups [64]. In addition, protein–polyphenol complexation may reduce IgE binding of allergens due to irreversible precipitation of allergens [16, 17]. A positive correlation was found between the strength of protein–polyphenol interactions and half-time of protein decay in gastric conditions and masking of total antioxidant capacity of protein–polyphenol complexes [65].

Interactions of polyphenols and proteins may result in precipitation, aggregation, and denaturation of proteins [62]. Polyphenols may directly influence digestive enzymes, and the resulting action may be enzyme inhibition or, in certain cases, enzyme activation [66, 67].

The effect of some phenolic compounds, resveratrol, catechin, EGCG, quercetin, and phenolics-rich beverages (red wine and green tea), on pepsin activity has been described previously. The tested polyphenols and beverages increased the initial velocity of the reaction, affecting the Vmax of pepsin on denatured hemoglobin as a substrate, and the activating effect is concentration dependent. [67]. On the contrary, many in vitro and in vivo studies showed antinutritive properties of polyphenols, especially tannins [68–70].

In vivo results showed antinutritive action of large polyphenols, i.e., tannins. The ability of polyphenolic compounds to form insoluble complexes with proteins has long been associated with the observed reduction in nutritive value resulting from their inclusion in animal diets. In addition, such complexation may reduce IgE binding of allergens due to irreversible precipitation, as recently shown for complexes of peanut allergens and phenolic acids [71]. Naturally occurring polyphenols, and in particular condensed tannins isolated from various plant sources, have been shown to inhibit in vitro a number of digestive enzymes including trypsin, alpha-amylase, and lipase [69, 72]. Covalent binding of phenolic acids to whey proteins adversely affects digestion by pepsin, trypsin, and hymotrypsin

[73]. It seems that a direct inhibition or activation of pepsin activity, but also interactions of various proteins with polyphenols, which to a different extent may bind to a formed network, are important factors in gastric digestion of complex protein/polyphenol networks. In different proteins, the same polyphenol molecule could have different binding affinities, which are related to the three-dimensional structure of the protein [74].

Solubility and aggregation of food allergens and larger peptides obtained during digestion seem to be an important allergy-promoting factor and directly linked to Th2 biasing of the immune response in animal models of food allergy [75].

Phenolics such as caffeic, chlorogenic, and ferulic acid readily form insoluble complexes with major peanut allergens, especially Ara h 1 and Ara h 2, and such complexation reduces IgE binding [71]. A recent study provided data on increasing binding affinities of oligomerized polyphenols to globular proteins [76].

7.3.6 Polyphenols and Intestinal Barrier Function

Impairment of the intercellular tight junction (TJ) shield, which is the major determinant of intestinal barrier function, is associated with various intestinal and metabolic diseases, such as inflammatory bowel disease, food allergy, and obesity. Intestinal TJ function is impaired in these diseases by various factors, such as inflammatory cytokines or ROS. Changes in epithelial cell morphology or adhesion may occur also due to proteolytic activity of dietary proteases, as demonstrated in vitro [77].

For four flavonoids—genistein, quercetin, myricetin, and green tea EGCG—it has been reported to exhibit protective and promotive effects on intestinal TJ barrier function (Fig. 7.3). Genistein and quercetin interact with intracellular signaling molecules, tyrosine kinases, and protein kinase C δ, resulting in the regulation of TJ protein expression and assembly [78, 79].

It has been shown that genistein, a major soybean isoflavone, protects TJ barrier function against oxidative stress, acetaldehyde, enteric bacteria, and inflammatory cytokines [80]. Genistein blocks the tyrosine phosphorylation of the TJ proteins induced by oxidative stress and acetaldehyde, which results in the disassembly of the proteins from the junctional complex. Quercetin, a flavonol, enhances intestinal TJ barrier function through the assembly and expression of TJ proteins. The change in phosphorylation status is responsible for the quercetin-mediated assembly of TJ proteins. TJ protein induction has an additional role in this effect. Myricetin, one of the flavonols found in grapes and tea, exhibits a promotive effect on intestinal TJ barrier function in Caco-2 cells. A green tea flavonoid, EGCG, was found to ameliorate the intestinal TJ barrier dysfunction provoked by interferon γ [77], but the mechanism has not been elucidated yet.

Future investigations are required to elucidate the precise mechanisms underlying these flavonoid-mediated protective effects on intestinal TJ barrier function, which may be beneficial in food allergy and other disorders in which intestinal barrier function is impaired.

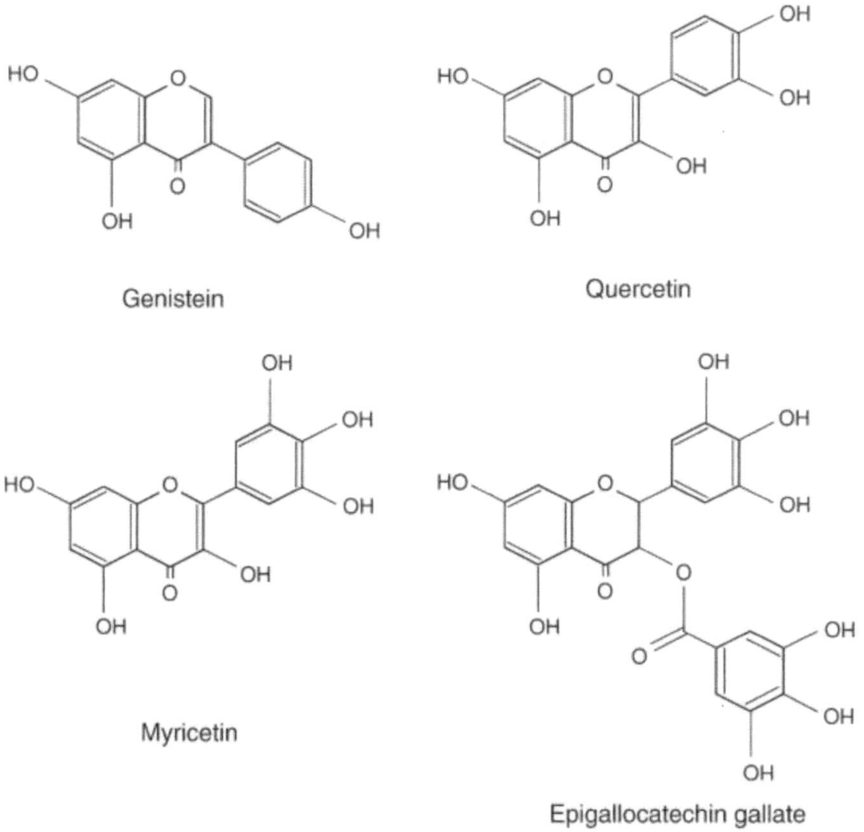

Fig. 7.3 Chemical structures of flavonoids active at the intestinal TJs barrier function: genistein, quercetin, myricetin, and EGCG

References

1. Devereux G, Seaton A: **Diet as a risk factor for atopy and asthma**. *J Allergy Clin Immunol* 2005, **115**(6):1109–1117; quiz 1118.
2. Kawai M, Hirano T, Higa S, Arimitsu J, Maruta M, Kuwahara Y, Ohkawara T, Hagihara K, Yamadori T, Shima Y *et al*: **Flavonoids and related compounds as anti-allergic substances**. *Allergol Int* 2007, **56**(2):113–123.
3. Bellik Y, Boukraa L, Alzahrani HA, Bakhotmah BA, Abdellah F, Hammoudi SM, Iguer-Ouada M: **Molecular mechanism underlying anti-inflammatory and anti-allergic activities of phytochemicals: an update**. *Molecules* 2012, **18**(1):322–353.
4. Liu RH: **Potential synergy of phytochemicals in cancer prevention: mechanism of action**. *J Nutr* 2004, **134**(12 Suppl):3479S–3485S.
5. Yang CS, Wang X, Lu G, Picinich SC: **Cancer prevention by tea: animal studies, molecular mechanisms and human relevance**. *Nat Rev Cancer* 2009, **9**(6):429–439.
6. Dou QP: **Molecular mechanisms of green tea polyphenols**. *Nutr Cancer* 2009, **61**(6):827–835.
7. Butt MS, Sultan MT: **Green tea: nature's defense against malignancies**. *Crit Rev Food Sci Nutr* 2009, **49**(5):463–473.

8. Teillet F, Boumendjel A, Boutonnat J, Ronot X: **Flavonoids as RTK inhibitors and potential anticancer agents.** *Med Res Rev* 2008, **28**(5):715–745.

9. Bode AM, Dong Z: **Epigallocatechin 3-gallate and green tea catechins: United they work, divided they fail.** *Cancer Prev Res (Phila Pa)* 2009, **2**(6):514–517.

10. Singh R, Akhtar N, Haqqi TM: **Green tea polyphenol epigallocatechin-3-gallate: inflammation and arthritis. [corrected].** *Life Sci* 2010, **86**(25–26):907–918.

11. Shahidi F, McDonald J, Chandrasekara A, Zhong Y: **Phytochemicals of foods, beverages and fruit vinegars: chemistry and health effects.** *Asia Pacific Journal of Clinical Nutrition* 2008, **17**:380–382.

12. McDougall GJ, Kulkarni NN, Stewart D: **Current developments on the inhibitory effects of berry polyphenols on digestive enzymes.** *Biofactors* 2008, **34**(1):73–80.

13. Lewis KC, Selzer T, Shahar C, Udi Y, Tworowski D, Sagi I: **Inhibition of pectin methyl esterase activity by green tea catechins.** *Phytochemistry* 2008, **69**(14):2586–2592.

14. Allan K, Kelly FJ, Devereux G: **Antioxidants and allergic disease: a case of too little or too much?** *Clin Exp Allergy* 2010, **40**(3):370–380.

15. Tanaka T, Kouda K, Kotani M, Takeuchi A, Tabei T, Masamoto Y, Nakamura H, Takigawa M, Suemura M, Takeuchi H *et al*: **Vegetarian diet ameliorates symptoms of atopic dermatitis through reduction of the number of peripheral eosinophils and of PGE2 synthesis by monocytes.** *J Physiol Anthropol Appl Human Sci* 2001, **20**(6):353–361.

16. Chung SY, Champagne ET: **Effects of Phytic acid on peanut allergens and allergenic properties of extracts.** *J Agric Food Chem* 2007, **55**(22):9054–9058.

17. Chung S, Champagne ET: **Effect of phytic acid on IgE binding to peanut allergens.** *J Allergy Clin Immunol* 2006, **117**(2):S38–S38.

18. Shim JH, Choi HS, Pugliese A, Lee SY, Chae JI, Choi BY, Bode AM, Dong Z: **(−)-Epigallocatechin gallate regulates CD3-mediated T cell receptor signaling in leukemia through the inhibition of ZAP-70 kinase.** *J Biol Chem* 2008, **283**(42):28370–28379.

19. Hirakawa S, Saito R, Ohara H, Okuyama R, Aiba S: **Dual Oxidase 1 Induced by Th2 Cytokines Promotes STAT6 Phosphorylation via Oxidative Inactivation of Protein Tyrosine Phosphatase 1B in Human Epidermal Keratinocytes.** *J Immunol* 2011, **186**(8):4762–4770.

20. Kawai K, Tsuno NH, Kitayama J, Okaji Y, Yazawa K, Asakage M, Hori N, Watanabe T, Takahashi K, Nagawa H: **Epigallocatechin gallate attenuates adhesion and migration of CD8+ T cells by binding to CD11b.** *J Allergy Clin Immunol* 2004, **113**(6):1211–1217.

21. Kawai K, Tsuno NH, Kitayama J, Okaji Y, Yazawa K, Asakage M, Hori N, Watanabe T, Takahashi K, Nagawa H: **Epigallocatechin gallate, the main component of tea polyphenol, binds to CD4 and interferes with gp120 binding.** *J Allergy Clin Immunol* 2003, **112**(5):951–957.

22. Umeda D, Yano S, Yamada K, Tachibana H: **Green tea polyphenol epigallocatechin-3-gallate signaling pathway through 67-kDa laminin receptor.** *J Biol Chem* 2008, **283**(6):3050–3058.

23. Van Aller GS, Carson JD, Tang W, Peng H, Zhao L, Copeland RA, Tummino PJ, Luo L: **Epigallocatechin gallate (EGCG), a major component of green tea, is a dual phosphoinositide-3-kinase/mTOR inhibitor.** *Biochem Biophys Res Commun* 2011, **406**(2):194–199.

24. Kawai K, Tsuno NH, Kitayama J, Okaji Y, Yazawa K, Asakage M, Sasaki S, Watanabe T, Takahashi K, Nagawa H: **Epigallocatechin gallate induces apoptosis of monocytes.** *J Allergy Clin Immunol* 2005, **115**(1):186–191.

25. Yoneyama S, Kawai K, Tsuno NH, Okaji Y, Asakage M, Tsuchiya T, Yamada J, Sunami E, Osada T, Kitayama J *et al*: **Epigallocatechin gallate affects human dendritic cell differentiation and maturation.** *J Allergy Clin Immunol* 2008, **121**(1):209–214.

26. Nakagawa K, Okuda S, Miyazawa T: **Dose-dependent incorporation of tea catechins, (−)-epigallocatechin-3-gallate and (−)-epigallocatechin, into human plasma.** *Biosci Biotechnol Biochem* 1997, **61**(12):1981–1985.

27. Ullmann U, Haller J, Decourt JP, Girault N, Girault J, Richard-Caudron AS, Pineau B, Weber P: **A single ascending dose study of epigallocatechin gallate in healthy volunteers.** *J Int Med Res* 2003, **31**(2):88–101.

28. Ahn SC, Kim GY, Kim JH, Baik SW, Han MK, Lee HJ, Moon DO, Lee CM, Kang JH, Kim BH et al: **Epigallocatechin-3-gallate, constituent of green tea, suppresses the LPS-induced phenotypic and functional maturation of murine dendritic cells through inhibition of mitogen-activated protein kinases and NF-kappaB**. *Biochem Biophys Res Commun* 2004, **313**(1):148–155.

29. Matsuo N, Yamada K, Shoji K, Mori M, Sugano M: **Effect of tea polyphenols on histamine release from rat basophilic leukemia (RBL-2H3) cells: the structure-inhibitory activity relationship**. *Allergy* 1997, **52**(1):58–64.

30. Fujimura Y, Umeda D, Yamada K, Tachibana H: **The impact of the 67 kDa laminin receptor on both cell-surface binding and anti-allergic action of tea catechins**. *Arch Biochem Biophys* 2008, **476**(2):133–138.

31. Inoue T, Suzuki Y, Ra C: **Epigallocatechin-3-gallate inhibits mast cell degranulation, leukotriene C(4) secretion, and calcium influx via mitochondrial calcium dysfunction**. *Free Radic Biol Med* 2010, **49**(4):632–640.

32. Wong CP, Nguyen LP, Noh SK, Bray TM, Bruno RS, Ho E: **Induction of regulatory T cells by green tea polyphenol EGCG**. *Immunol Lett* 2011, **139**(1–2):7–13.

33. Melgarejo E, Medina MA, Sanchez-Jimenez F, Urdiales JL: **Epigallocatechin gallate reduces human monocyte mobility and adhesion in vitro**. *Br J Pharmacol* 2009, **158**(7):1705–1712.

34. Danesi F, Philpott M, Huebner C, Bordoni A, Ferguson LR: **Food-derived bioactives as potential regulators of the IL-12/IL-23 pathway implicated in inflammatory bowel diseases**. *Mutat Res* 2010, **690**(1–2):139–144.

35. Wu D, Guo Z, Ren Z, Guo W, Meydani SN: **Green tea EGCG suppresses T cell proliferation through impairment of IL-2/IL-2 receptor signaling**. *Free Radic Biol Med* 2009, **47**(5):636–643.

36. Pae M, Ren Z, Meydani M, Shang F, Meydani SN, Wu D: **Epigallocatechin-3-gallate directly suppresses T cell proliferation through impaired IL-2 utilization and cell cycle progression**. *J Nutr* 2010, **140**(8):1509–1515.

37. Hushmendy S, Jayakumar L, Hahn AB, Bhoiwala D, Bhoiwala DL, Crawford DR: **Select phytochemicals suppress human T-lymphocytes and mouse splenocytes suggesting their use in autoimmunity and transplantation**. *Nutr Res* 2009, **29**(8):568–578.

38. Noh SU, Cho EA, Kim HO, Park YM: **Epigallocatechin-3-gallate improves Dermatophagoides pteronissinus extract-induced atopic dermatitis-like skin lesions in NC/Nga mice by suppressing macrophage migration inhibitory factor**. *Int Immunopharmacol* 2008, **8**(9):1172–1182.

39. Wang J, Pae M, Meydani SN, Wu D: **Epigallocatechin-3-gallate inhibits expression of receptors for T cell regulatory cytokines and their downstream signaling in mouse CD4+ T cells**. *J Nutr* 2012, **142**(3):566–571.

40. Wu H, Zhu B, Shimoishi Y, Murata Y, Nakamura Y: **(−)-Epigallocatechin-3-gallate induces up-regulation of Th1 and Th2 cytokine genes in Jurkat T cells**. *Arch Biochem Biophys* 2009, **483**(1):99–105.

41. Luz Sanz M, Corzo-Martinez M, Rastall RA, Olano A, Moreno FJ: **Characterization and in vitro digestibility of bovine beta-lactoglobulin glycated with galactooligosaccharides**. *J Agric Food Chem* 2007, **55**(19):7916–7925.

42. Huang CH, Jan RL, Kuo CH, Chu YT, Wang WL, Lee MS, Chen HN, Hung CH: **Natural flavone kaempferol suppresses chemokines expression in human monocyte THP-1 cells through MAPK pathways**. *J Food Sci* 2010, **75**(8):H254–259.

43. Lee EJ, Ji GE, Sung MK: **Quercetin and kaempferol suppress immunoglobulin E-mediated allergic inflammation in RBL-2H3 and Caco-2 cells**. *Inflamm Res* 2010, **59**(10):847–854.

44. Gong JH, Shin D, Han SY, Kim JL, Kang YH: **Kaempferol suppresses eosionphil infiltration and airway inflammation in airway epithelial cells and in mice with allergic asthma**. *J Nutr* 2012, **142**(1):47–56.

45. O'Shea JJ, Plenge R: **JAK and STAT signaling molecules in immunoregulation and immune-mediated disease**. *Immunity* 2012, **36**(4):542–550.

46. Nelms K, Keegan AD, Zamorano J, Ryan JJ, Paul WE: **The IL-4 receptor: signaling mechanisms and biologic functions**. *Annu Rev Immunol* 1999, **17**:701–738.
47. Cortes JR, Perez-G M, Rivas MD, Zamorano J: **Kaempferol inhibits IL-4-Induced STAT6 activation by specifically targeting JAK3**. *J Immunol* 2007, **179**(6):3881–3887.
48. Yamashita S, Yamashita T, Yamada K, Tachibana H: **Flavones suppress type I IL-4 receptor signaling by down-regulating the expression of common gamma chain**. *Febs Letters* 2010, **584**(4):775–779.
49. Higa S, Hirano T, Kotani M, Matsumoto M, Fujita A, Suemura M, Kawase I, Tanaka T: **Fisetin, a flavonol, inhibits TH2-type cytokine production by activated human basophils**. *J Allergy Clin Immunol* 2003, **111**(6):1299–1306.
50. Hirano T, Higa S, Arimitsu J, Naka T, Shima Y, Ohshima S, Fujimoto M, Yamadori T, Kawase I, Tanaka T: **Flavonoids such as luteolin, fisetin and apigenin are inhibitors of interleukin-4 and interleukin-13 production by activated human basophils**. *Int Arch Allergy Immunol* 2004, **134**(2):135–140.
51. Huang RY, Yu YL, Cheng WC, OuYang CN, Fu E, Chu CL: **Immunosuppressive effect of quercetin on dendritic cell activation and function**. *J Immunol* 2010, **184**(12):6815–6821.
52. Okada Y, Oh-oka K, Nakamura Y, Ishimaru K, Matsuoka S, Okumura K, Ogawa H, Hisamoto M, Okuda T, Nakao A: **Dietary resveratrol prevents the development of food allergy in mice**. *PLoS One* 2012, **7**(9):e44338.
53. Fewtrell CM, Gomperts BD: **Effect of flavone inhibitors of transport ATPases on histamine secretion from rat mast cells**. *Nature* 1977, **265**(5595):635–636.
54. Pearce FL, Befus AD, Bienenstock J: **Mucosal mast cells. III. Effect of quercetin and other flavonoids on antigen-induced histamine secretion from rat intestinal mast cells**. *J Allergy Clin Immunol* 1984, **73**(6):819–823.
55. Middleton E, Jr., Drzewiecki G, Krishnarao D: **Quercetin: an inhibitor of antigen-induced human basophil histamine release**. *J Immunol* 1981, **127**(2):546–550.
56. Cheong H, Ryu SY, Oak MH, Cheon SH, Yoo GS, Kim KM: **Studies of structure activity relationship of flavonoids for the anti-allergic actions**. *Arch Pharm Res* 1998, **21**(4):478–480.
57. Lee TP, Matteliano ML, Middleton E, Jr.: **Effect of quercetin on human polymorphonuclear leukocyte lysosomal enzyme release and phospholipid metabolism**. *Life Sci* 1982, **31**(24):2765–2774.
58. Lambert JD, Elias RJ: **The antioxidant and pro-oxidant activities of green tea polyphenols: A role in cancer prevention**. *Arch Biochem Biophys* 2010, **501**(1):65–72.
59. Kaminski MM, Sauer SW, Klemke CD, Suss D, Okun JG, Krammer PH, Gulow K: **Mitochondrial reactive oxygen species control T cell activation by regulating IL-2 and IL-4 expression: mechanism of ciprofloxacin-mediated immunosuppression**. *J Immunol* 2010, **184**(9):4827–4841.
60. Naik E, Dixit VM: **Mitochondrial reactive oxygen species drive proinflammatory cytokine production**. *J Exp Med* 2011, **208**(3):417–420.
61. Bulua AC, Simon A, Maddipati R, Pelletier M, Park H, Kim KY, Sack MN, Kastner DL, Siegel RM: **Mitochondrial reactive oxygen species promote production of proinflammatory cytokines and are elevated in TNFR1-associated periodic syndrome (TRAPS)**. *J Exp Med* 2011, **208**(3):519–533.
62. Bandyopadhyay P, Ghosh AK, Ghosh C: **Recent developments on polyphenol-protein interactions: effects on tea and coffee taste, antioxidant properties and the digestive system**. *Food Funct* 2012, **3**(6):592–605.
63. Cao H, Shi Y, Chen X: **Advances on the interaction between tea catechins and plasma proteins: structure-affinity relationship, influence on antioxidant activity, and molecular docking aspects**. *Curr Drug Metab* 2013, **14**(4):446–450.
64. Bras NF, Goncalves R, Mateus N, Fernandes PA, Ramos MJ, Do Freitas V: **Inhibition of Pancreatic Elastase by Polyphenolic Compounds**. *J Agric Food Chem* 2010, **58**(19):10668–10676.

65. Stojadinovic M, Radosavljevic J, Ognjenovic J, Vesic J, Prodic I, Stanic-Vucinic D, Cirkovic Velickovic T: **Binding affinity between dietary polyphenols and beta-lactoglobulin negatively correlates with the protein susceptibility to digestion and total antioxidant activity of complexes formed.** *Food Chem* 2013, **136**(3–4):1263–1271.

66. Tantoush Z, Apostolovic D, Kravic B, Prodic I, Mihajlovic L, Stanic-Vucinic D, Cirkovic Velickovic T: **Green tea catechins of food, supplements facilitate pepsin digestion of major food allergens, but hampers their digestion if oxidized by phenol oxidase.** *J Funct Food* 2012, **4**(3):650–660.

67. Tagliazucchi D, Verzelloni E, Conte A: **Effect of some phenolic compounds and beverages on pepsin activity during simulated gastric digestion.** *J Agric Food Chem* 2005, **53**(22):8706–8713.

68. Bravo L: **Polyphenols: chemistry, dietary sources, metabolism, and nutritional significance.** *Nutr Rev* 1998, **56**(11):317–333.

69. Grussu D, Stewart D, McDougall GJ: **Berry Polyphenols Inhibit alpha-Amylase in Vitro: Identifying Active Components in Rowanberry and Raspberry.** *J Agric Food Chem* 2011, **59**(6):2324–2331.

70. Goncalves B, Landbo AK, Knudsen D, Silva AP, Moutinho-Pereira J, Rosa E, Meyer AS: **Effect of ripeness and postharvest storage on the phenolic profiles of Cherries (Prunus avium L.).** *J Agric Food Chem* 2004, **52**(3):523–530.

71. Chung S-Y, Champagne ET: **Reducing the allergenic capacity of peanut extracts and liquid peanut butter by phenolic compounds.** *Food Chemistry* 2009, **115**(4):1345–1349.

72. Griffiths DW: **The inhibition of digestive enzymes by polyphenolic compounds.** *Adv Exp Med Biol* 1986, **199**:509–516.

73. Rawel HM, Kroll J, Hohl UC: **Model studies on reactions of plant phenols with whey proteins.** *Nahrung* 2001, **45**(2):72–81.

74. Soares S, Mateus N, Freitas V: **Interaction of different polyphenols with bovine serum albumin (BSA) and human salivary alpha-amylase (HSA) by fluorescence quenching.** *J Agric Food Chem* 2007, **55**(16):6726–6735.

75. Roth-Walter F, Berin MC, Arnaboldi P, Escalante CR, Dahan S, Rauch J, Jensen-Jarolim E, Mayer L: **Pasteurization of milk proteins promotes allergic sensitization by enhancing uptake through Peyer's patches.** *Allergy* 2008, **63**(7):882–890.

76. Prigent SV, Voragen AG, van Koningsveld GA, Baron A, Renard CM, Gruppen H: **Interactions between globular proteins and procyanidins of different degrees of polymerization.** *J Dairy Sci* 2009, **92**(12):5843–5853.

77. Cavic M, Grozdanovic M, Bajic A, Srdic-Rajic T, Andjus PR, Gavrovic-Jankulovic M: **Actinidin, a protease from kiwifruit, induces changes in morphology and adhesion of T84 intestinal epithelial cells.** *Phytochemistry* 2012, **77**:46–52.

78. Suzuki T, Hara H: **Role of flavonoids in intestinal tight junction regulation.** *J Nutr Biochem* 2011, **22**(5):401–408.

79. Suzuki T, Hara H: **Quercetin enhances intestinal barrier function through the assembly of zonula [corrected] occludens-2, occludin, and claudin-1 and the expression of claudin-4 in Caco-2 cells.** *J Nutr* 2009, **139**(5):965–974.

80. Atkinson KJ, Rao RK: **Role of protein tyrosine phosphorylation in acetaldehyde-induced disruption of epithelial tight junctions.** *Am J Physiol Gastrointest Liver Physiol* 2001, **280**(6):G1280–1288.

Chapter 8
Predicting Potential Allergenicity of New Proteins Introduced by Biotechnology

Contents

Abbreviations

ARP	Allergen-representative peptides
BLAST	Basic local alignment search tool
DASARP	Detection based on automated selection of allergen-representative peptides
DDBJ	DNA Data Bank of Japan
EMBL	European molecular biology laboratory
ELISA	Enzyme-linked immunosorbent assay
FAO	Food and agriculture organization
GM	Genetically modified
HMM	Hidden Markov model
MEME	Motif-based sequence analysis
NCBI	National center for biotechnology information
PAGE	Polyacrylamide gel electrophoresis
PD	Propensity distance
QSAR	Quantitative structure-activity relationship
SDS	Sodium dodecyl sulfate
SGF	Simulated gastric fluid
SEB	Staphylococcal enterotoxin B
SDAP	Structural database of allergenic proteins
WHO	World Health Organization

T. Ćirković Veličković, M. Gavrović-Jankulović, *Food Allergens,*
Food Microbiology and Food Safety, DOI 10.1007/978-1-4939-0841-7_8,
© Springer Science+Business Media New York 2014

Summary The potential allergenicity of newly introduced proteins in genetically engineered foods has become an important safety evaluation issue. Food allergy is an important and common health issue, and therefore there is a need to characterize the sensitizing potential of novel food proteins. Approaches currently used include consideration of structural similarity to, or amino acid sequence homology with, known allergens using bioinformatics tools; immunologic cross-reactivity with known allergens; and the measurement of resistance to proteolytic digestion by pepsin in a simulated gastric fluid. Although these methods provide information that contributes to safety assessment, they do not provide a direct evaluation of the ability of a novel protein to cause allergic sensitization. For this reason, considerable interest exists in the design and evaluation of suitable animal models that may provide a more holistic assessment of allergenic potential. An appropriate animal model should produce sensitization and/or elicitation of allergic symptoms at a physiologically relevant dose, via the relevant route of exposure in a standard mouse strain. So far, developed mouse models of food allergy mostly use adjuvants (such as cholera toxin and staphylococcal enterotoxin B) and the oral route of exposure. None of the currently studied models has been widely accepted and validated. More work is needed on identification of appropriate end points, particularly those that reflect anaphylactic activity. Before validation can be considered, decisions have to be made regarding which mouse strains and adjuvants to include, as well as the doses of test materials. Appropriate test substances that represent a range from highly allergenic to poorly allergenic need to be selected. The data also indicate that the food matrix can influence responses to individual proteins and, therefore, the food matrix should be taken into account when developing models for predicting the allergenic potential of new proteins introduced by biotechnology.

8.1   Genetically Modified Food

Genetically modified (GM) plants are produced by altering the DNA of the plant genome through introduction, rearrangement, or removal of DNA. The resulting plants are commonly known as genetically engineered or GM plants; when used as food sources they are known as GM plant foods or GM foods. The resulting GM crops offer improved pest and disease resistance; higher yields; superior flavor, appearance, and nutrition; tolerance of specific herbicides; and reduced requirements for fertilizer or water [1, 2]. In addition, genetic improvement in agriculturally important plants has contributed to increased food, fiber, and energy production for centuries and increasingly so during the past 40 years [3]. Despite the benefits of GM crops, the potential health hazards of each genetically transformed food, including the risk of allergenicity, need to be carefully evaluated because food crops in common use are generally recognized as safe, except for individuals with specific food allergies. The primary concern is the potential transfer of a major allergen from a different species into a food crop, as was the case when a Brazil nut 2S albumin was transferred into soybean to improve nutritional quality [4]. This caution has led to strategies to monitor transformed crops for allergenic potential before release. Such approaches are based on what is known about the pathogenesis of food allergy and the characteristics of food allergens. In the process, immunologists, allergists, and food technologists have been challenged to translate scientific and clinical observations into practical approaches to prevent the creation and marketing of new allergenic foods [2].

Allergenicity is not an intrinsic, fully predictable property of a given protein but is a biological activity requiring an interaction with individuals with a predisposed genetic background. Allergenicity therefore depends upon the genetic diversity and variability in atopic humans. Frequency, severity, and specificity of allergic reactions also depend upon geographic and environmental factors. Given this lack of complete predictability, it is necessary to consider several aspects in the risk assessment process, to obtain a cumulative body of evidence which minimizes any uncertainty with regard to the protein(s) in question [5, 6].

New GM crop requires a premarket safety assessment to evaluate intended and unintended changes that might have adverse human health consequences caused by the transfer of the deoxyribonucleic acid (DNA, genes). The goal is to identify hazards, and if found, to require risk assessment and where appropriate develop a risk management strategy.

The safety assessment is based on scientific observations and requires the use of methods and criteria that are demonstrated to be predictive [7]. The framework to guide the evaluation of potential safety issues requires the following characteristics:

- The GM plant and its use as food
- The source of the gene
- The inserted DNA and flanking DNA at the insertion site
- The expressed substances (e.g., proteins and any new metabolites that result from the new gene product)
- The potential toxicity and antinutritional properties of new proteins or metabolites
- The introduced protein compared with those known to cause celiac disease if the DNA is from wheat, barley, rye, oats, or related grains
- The introduced protein for potential allergenicity
- Key endogenous nutrients and antinutrients, including toxins and allergens, for potential increases for specific host plants (DNA recipients) [8]

The first document related to the guidelines for allergenicity assessment of GM crops was published in 1996 by the International Food Biotechnology Council (IFBC, Washington, DC) in collaboration with the International Life Sciences Institute (ILSI, Washington, DC) [9]. This document was followed by the UN Food and Agriculture Organization (FAO)/World Health Organization (WHO) consultation recommendations in 2001 [10]; then, in 2003, by the Codex Alimentarius Commission guidelines [6]; and finally, in 2009, by recommendations for the foods derived from modern biotechnology [7].

The evaluation process begins with an evaluation of the *source of the gene*. If the source of the gene encoding the new protein is a commonly allergenic food (e.g., peanut, hazelnut, hen's egg, or cow's milk), IgE-binding studies using sera from patients allergic to the source are required to ensure that the encoded protein does not bind IgE from those allergic to the source. For serum selection, demographic factors need to be taken into account together with the number of sera included in the evaluation procedure.

When the introduced genetic material is obtained from wheat, rye, barley, oats, or related cereal grains, the applicant should also assess the newly expressed proteins for a possible role in the elicitation of gluten-sensitive enteropathy or other enteropathies which are not IgE-mediated. Where events have been stacked, the applicant should provide an assessment of any potential for increased allergenicity to humans and animals on a case-by-case approach. These potential effects may arise from additive, synergistic, or antagonistic effects of the gene products [5].

Bioinformatics Analysis The amino acid sequence of all transferred proteins should be compared with known allergens by FASTA or Basic Local Alignment Search Tool (BLAST) algorithms to determine if any identity match is sufficiently high to suspect that the protein might cause allergic cross-reactions. Such analysis should identify proteins that would require serum testing, using donors with specific allergies to evaluate potential IgE binding. If the sequence identity match is high (e.g., >70% over most of the length of the protein), then the potential for cross-reactivity is also high. Matches sharing between 50 and 70% of the sequence identity pose a moderate risk of cross-reactivity and should be tested for IgE binding. If the match is <50% identical, the risk of cross-reactivity is expected to be low [10]. A threshold value of 35% identity over any 80-amino-acid segment of the transferred protein is defined in both the FAO/WHO [8] and Codex documents [6]. It was introduced to identify conserved gene segments representing functional motifs, which might retain the conformational epitope structure as well. Proteins with higher matching identities (e.g., >35% identity) are recommended for testing of IgE binding [8].

IgE Reactivity When there is an indication of sequence homology or structure similarities, an important procedure for assessing the potential that exposure to the newly expressed proteins might elicit an allergic reaction in individuals already sensitized to cross-reactive proteins, is based on in vitro tests that measure the capacity of specific IgE from serum of allergic patients to bind the test protein(s). It is noted that there is interindividual variability in the specificity and affinity of the human IgE response. In particular, the specificity of the IgE antibodies to the different allergens present in a given food/source and/or to the different epitopes present on a given protein may vary among allergic individuals. In order to optimize the sensitivity of the test, individual sera from well-characterized allergic individuals should be used rather than pooled sera [5]. Serum IgE testing to evaluate proteins from an allergenic source, or proteins with sequence identity (e.g., >35% over an 80-amino-acid window or >50% overall) to a known allergen, should be capable of detecting IgE binding to linear and conformational epitopes. Potential difficulties in interpretations of the results could be the molecular appearance of the protein (correct protein folding, presence of disulfide bonds, and presence of N-linked glycans).

Serum from individuals with carbohydrate-specific IgE should be avoided for GM assessment, and selection of donors with IgE directed against peptide epitopes rather than carbohydrate moieties is recommended. Carbohydrate-binding sera would lead to detection of glycoproteins as an allergenic risk, although there is a consensus in the scientific community that the glycans are unlikely to cause clinical food allergy [11, 12].

Stability to Proteolytic Enzymes Stability to digestion by proteolytic enzymes has long been considered a characteristic of allergenic proteins. Some potent food allergens are known to be stable in an in vitro pepsin digestion assay, whereas most of dietary proteins are readily digestible [13]. Although it has been established that no absolute correlation exists [14], resistance of proteins to pepsin digestion is still proposed as an additional criterion to be considered in an overall risk assessment. The pepsin resistance test is generally performed under quite standardized conditions [15], at low pH values and high pepsin-to-protein ratios, although the pepsin resistance test does not reflect the physiological conditions of the digestion. In addition, the digestibility of the newly expressed proteins in specific segments of the population such as infants and individuals with impaired digestive functions may be assessed using in vitro digestibility tests under different conditions [16]. Also, since the protein encoded by the newly introduced genes will be present in the product as a complex matrix, the impact of the possible interaction between the protein and other components of the matrix, as well as the effects of the processing, should be taken into account in additional in vitro digestibility tests. Depending on the outcome of the in vitro digestibility test, it could also be useful to compare intact, heat-denatured, and pepsin-digested proteins for IgE binding, since an altered digestibility may impact on the allergenicity of the newly expressed protein.

Although FAO/WHO recommendations called for evaluation of new GM crop with studies in two separate species of animals and/or using two routes of sensitizations in one species, it seems that no validated animal model is predictive of allergenicity to food proteins in humans [8].

In 2009, the FAO/WHO recommendations for foods derived from modern biotechnology were launched. The term "modern biotechnology" actually encompasses the application of: (1) in vitro nucleic acid techniques, including recombinant DNA and direct injection of nucleic acid into cells or organelles; or (2) fusion of cells beyond the taxonomic family that overcome natural physiological reproductive or recombinant barriers and that are not techniques used in traditional breeding and selection [7].

Since there is no definitive test that can be relied upon to predict allergic response in humans to a newly expressed protein, it is recommended that an integrated, stepwise, case-by-case approach, be used in the assessment of possible allergenicity of newly expressed proteins. Besides guidelines given in the previous directives, the nature of the food product intended for consumption should be taken into consideration in determining the types of processing that would be applied and its effects on the presence of the protein in the final food product. As scientific knowledge and technology evolve, other methods and tools may be considered in assessing the allergenicity potential of newly expressed proteins as part of the assessment strategy, such as targeted serum screening (i.e., the assessment of binding to IgE in sera of individuals with clinically validated allergic responses to broadly related categories of foods); the development of international serum banks; use of animal models; and examination of newly expressed proteins for T cell epitopes and structural motifs associated with allergens [5].

8.2   Homology and Structural Similarity: Bioinformatics of Food Allergens

The early discoveries in the 1960s indicating that protein sequences bear information about their ultimate structure and function, together with the relative simplicity with which these sequences could be obtained in the laboratory, were impetus for trying to predict the structure and function from a sequence using computer-based methods, i.e., bioinformatics. The origins of bioinformatics lie in the field of structural biology, because many of the first bioinformatics programs and databases were developed to store, compare, and analyze protein structures [17]. The pioneer work in collection of all known amino acid sequences of that time was the *Atlas of Protein Sequence and Structure,* prepared by Margaret Oakley Dayhoff with her coworkers at the Protein Information Resource, which was published in 1965. Nowadays, different methods use different algorithms to store, compare, and analyze protein sequences; however, the logic behind them is fairly similar. Most of these algorithms belong to a class of approaches that is called machine-learning algorithms, which offer different ways to learn what the characteristic features of the entity of interest are, based on a large number of examples. Much of bioinformatics today is based on using computers to manipulate, store, and compare sequences or character strings [18], but besides that, it also deals with protein structure analysis—structural bioinformatics.

Approximately 90% of all food allergies are associated with a small number of specific proteins represented by eight major allergenic foods: peanuts, tree nuts, cow's milk, hen's eggs, fish, crustacean, wheat, and soybeans [9, 19]. The huge progress in deciphering the primary structure of proteins resulted in filling out various protein databases including the Pfam database, created by Wellcome trust Sanger Institute, which represents a large collection of protein sequences categorized into families. All allergenic proteins deposited in the Structural Database of Allergenic Proteins (SDAP) could be grouped to 130 (of 9,318 total) Pfams, and 31 families contain more than four allergens [20]. Radauer et al. built the AllFam database of allergen families (http://www.meduniwien.ac.at/allergens/allfam/) and employed it to extract common structural and functional properties of allergens. Seven hundred and seven allergens were classified by sequence into 134 AllFam families containing 184 Pfam domains (2% of 9,318 Pfam families) [21]. However, despite the great number of identified allergenic proteins [22], it is still not known why a relatively small number of certain proteins provoke allergenic reactions in humans. In this regard, the development of a method for allergenicity prediction would be beneficial, especially in order to prevent the inadvertent generation of new allergenic food plants by agricultural biotechnology [23].

Over the past two decades, various approaches for identifying potential food allergens for purposes of safety assessment of genetically engineered crops have been developed and modified in order to better understand what characteristics make a protein an allergen [24]. In an attempt to design bioinformatic tools which can help in prediction of allergenic reactivity, various methodologies have been developed. For example, several computer algorithms now exist for discovering multiple motifs (expressed as weight matrices) that characterize a family of protein sequences known to be homologous. Stadler and Stadler defined 52 allergen motifs by comparing allergens to nonallergens using motif-based sequence analysis (MEME) tools [23], the algorithm employed for the discovery of protein sequence motifs [25]. Li et al. tried to identify allergenic motifs by clustering known allergens, followed by wavelet analysis and hidden Markov model (HMM) profile preparation of each identified motif [26]. Bjorklund et al. developed a detection method for allergen-representative peptides (ARPs) with low or no occurrence in proteins lacking allergenic properties. The method has been designated as detection based on automated selection of ARP (DASARP) and outperforms the criterion based on identical peptide match for predicting allergenicity recommended by ILSI/IFBC and FAO/WHO [27]. AlgPred has been developed for the prediction of allergenic proteins and for mapping IgE epitopes on allergenic proteins (http://www.imtech. res.in/raghava/algpred/) [28]. The AlgPred server for allergenic protein prediction combines several methods for allergen protein prediction: Support Vector Machines (SVM), MEME/MAST (MAST determines the best match in the sequence to each motif) [29], IgE epitopes, and ARP.

In opposition to programs based on sequence similarities, AllerTOP is an alignment-independent server for in silico prediction of allergens based on the main physicochemical properties of proteins. The amino acids in the protein sequences are described by three z-descriptors (z_1, z_2, and z_3) and by auto- and cross-covariance (ACC) transformation for conversion of proteins into uniform vectors. The descriptor z_1 reflects the hydrophobicity of amino acids, the descriptor z_2 reflects their size, and the descriptor z_3 their polarity. ACC is a protein sequence analysis method developed by Wold and colleagues [30], which has been applied to quantitative structure–activity relationship (QSAR) studies of peptides with different length [31] and for protein classification [32]. The ACC transformation accounts for neighbor effects, i.e., the lack of independence between different sequence positions [33].

8.2.1 Allergen Databases

In the past years, significant improvements have occurred in the understanding of how to use bioinformatic analysis as part of an allergenicity risk assessment process for bioengineered foods. The efficacy of any specific bioinformatic analysis of allergenic proteins is dependent on the nature of the data sets that are employed in the analysis. A number of different allergen-related databases have been developed, which differ in the level of annotation, whether there are linkages to the repository databases that are the sources for sequence and structural data [34]. Two basic types of food allergen data resources have been developed: those that primarily provide clinical, physiological, or epidemiological information on food allergy and those that primarily provide molecular information for allergenic proteins.

Starting from the first release of the International Union of Immunological Societies (IUIS) Allergen Nomenclature subcommittee website [35], several research groups and institutions have started to accumulate information and data from available sources in order to create databases of scientific knowledge on allergens. A number of allergen databases are now accessible. An overview on the most updated list of allergen-dedicated Web-based resources is given in a study by Mari et al. [36]. There are several other databases available supporting computational tools [23, 37] or from web sites of companies involved in the allergy field.

An important database that does not fit into these categories is the Allergen Nomenclature database of the IUIS Allergen Nomenclature Sub-Committee (http://www.allergen.org). The database is intended to provide a central resource for insuring that allergen designations are uniform and consistent [38–40]. Critical feature of the allergen naming process is the requirement for clinical information demonstrating allergenic activity.

The enormous amount of scientific data generated during the past decades led to an exponential growth of biological databases. Among the primary biological databases, the UniProtKB/Swiss-Prot knowledgebase is a central resource for protein sequences and functional information [41], while GenBank/EMBL/DDBJ represent annotated collections of all publicly available DNA sequences. The RCSB Protein Data Bank encompasses all the experimentally determined structures of proteins, DNA, and other complex assemblies [42].

The Biotechnology Information for Food Safety database, released in 1998 by the National Center for Food Safety and Technology in Chicago, was the first attempt to provide a complete, nonredundant list of allergens from food and nonfood sources, with the initial aim of assessing the potential allergenicity of GM foods [22]. Allergen source organisms were linked to the National Center for Biotechnology Information (NCBI) taxonomy database [43].

AllergenOnline (http://www.allergenonline.org) was established in 2002 within the Food Allergy Research and Resource Program at the University of Nebraska [44]. AllergenOnline version 14 from January 2014 covers 1,706 peer-reviewed sequences, including 645 taxonomic protein groups. Features and tools available for database search are: (1) for full-length alignments by FASTA, as the most predictive search is the overall FASTA alignment, with identity matches greater than 50 % indicating possible cross-reactivity); (2) search for 80 amino acid alignments by FASTA, a precautionary search using a sliding window of 80 amino acid segments of each protein to find identities greater than 35 % (according to Codex Alimentarius guidelines, 2003); (3) search for an eight-amino-acid exact match. A panel of scientists and clinicians are involved in reviewing data for inclusion of proteins in the database by comparing peer-reviewed publications supporting the classification of the proteins as allergens or putative allergens following predetermined guidelines.

The Structural Database of Allergen Proteins (SDAP, http://fermi.utmb.edu/SDAP) was released in 2001 [45]. SDAP is a platform formed by a continuously updated allergen database. To date, SDAP database encompasses 1,526 allergens and isoallergens, 1,312 protein sequences for allergens and isoallergens, 92 allergens with PDB structures, 458 3D models for allergens and isoallergens, 29 allergens with IgE epitope sets, and 130 Pfam allergen classes. It provides computational tools for predicting the allergenicity of proteins and for studying the cross-reactivity between allergens and epitope search. The SDAP (University of Texas Medical Branch) is a Web server that integrates a database of allergenic proteins with various computational tools that can assist structural biology studies related to allergens. SDAP is a useful tool in the investigation of the cross-reactivity between known allergens, in testing the FAO/WHO allergenicity rules for new proteins, and in predicting the IgE-binding potential of GM food proteins. It is possible to retrieve information related to an allergen from the common protein sequence and structure databases, SwissProt [46], PIR [47], NCBI [48], and PDB [49], to find sequence and structural neighbors for an allergen and to search for the presence of an epitope other than the whole collection of allergens. SDAP can be used to determine food sources that might contain cross-reacting antigens. It also includes a peptide-matching function and a peptide similarity search based on a "propensity distance" (PD) value calculated from the physicochemical properties of the amino acids in a peptide. Besides SDAP, Allermatch [50] and AllerTool [51] also contain extensive database of known allergen proteins and use them in sequence searches of the query protein.

The Allergome (http://www.allergome.org) platform was released on the Web in 2003 with the aim of classifying allergens, IgE-binding antigens, and non-IgE-binding structures [36]. As of February 2014, it contained 2,261 allergen sources, with 7,131 links to molecule sequences, and a total of 27,094 bibliographic references. Allergome is updated daily on the basis of published literature. The criteria rely on structural relationships with known allergens and IgE-binding capacity to collect as many molecules as possible. Allergome contains all IUIS allergens, which are marked to distinguish them from non-IUIS structures, thus aiding in archive searches.

8.3 Testing Food Allergens Digestion

The digestive process plays an important role in the development of allergic sensitization, as well as in the clinical severity of food allergy symptoms since exposure of the immune system to proteins is required to initiate an allergic response. The gold standard for investigating food digestion is the use of *in vivo* approaches. Due to ethical concerns, animal studies are mostly performed. Development of *in vitro* models that mimic *in vivo* models are focused on three main stages: processing in

Table 8.1 An overview of the most common simulated gastric fluid (SGF) protocols used for testing food allergens digestibility

Investigator	SGF	pH	Pepsin concentration	Test protein	Pepsin to protein ratio	Assessment of the end point of digestion
Astwood et al. [53]	30 mM NaCl	1.2	0.32 % (3.2 mg/mL)	170 µg/mL of protein in 200 µL SGF	19:1	10–20 % SDS-PAGE
Fu et al. [14]	30 mM NaCL	1.2	0.32 % (3.2 mg/mL)	10 µL of the 5 mg/mL test protein were added to 200 µL of SGF	13:1	10–20 % Tris–Tricine PAGE
Thomas et al. [15] Ofori-Anti et al. [54]	0.084 N HCl, 35 mM NaCl	1.2 or 2.0	4,000 U in 1.52 mL of fluid (Sigma Pepsin was used, 3,460 U/mg of protein)	10 U of pepsin activity/µg of test protein was used in the assay; 0.08 ml of test protein solution (5 mg/ml) was added to 1.52 mL of SGF	3:1	10–20 % Tris–Tricine PAGE 10–20 % SDS-PAGE

the mouth, stomach, and duodenum. Static and dynamic *in vitro* models are being investigated. Static models are defined as those where products of digestion are not removed and which do not mimic the physical processes that occur *in vivo* (e.g., mechanical breakdown of food tissue, shearing, etc.). Dynamic models include both the physical processing and temporal changes in luminal conditions that mimic *in vivo* conditions. These models can use more complex foods and investigate the effects of the food matrix on the kinetics of food allergen release [52].

For traditional reasons, parameters of digestion stability of a food allergen are usually tested in a simple, static assay, in which pH of the gastric fluid varies from 1.2 to 2.5, with the end point of digestion and protein stability assessed usually by polyacrylamide gel electrophoresis (PAGE; Table 8.1).

The first published application of an *in vitro* pepsin digestion assay to address the question of food allergen stability was by Astwood et al. [53]. In that study, a number of food allergens or their peptide fragments were shown to be resistant to digestive proteolysis *in vitro*. Since this initial report, there have been several studies repeating the pepsin digestion assay for a variety of proteins.

Digestion stability of a protein depends on the composition of the digestion fluid and pepsin concentration and activity, parameters that often varied among studies conducted on the food allergens digestibility, thus unavoidably leading to inconsistency of these data in the literature.

The disagreements in literature may largely be attributed to variations in the amount and purity of enzymes or a protein used in the assay, pepsin-to-test protein ratio, pH value of the SGF, temperature of the reaction, or methods of end point detection. Assessing the end point of digestion by electrophoresis can also lead to inconsistency of the data due to variations in laboratory protocols used for electrophoretic gels fixation and staining [15].

Thomas et al. assessed the pepsin digestibility of a common set of proteins in nine independent laboratories to determine the reproducibility of the assay when performed using a common protocol. Results were relatively consistent across laboratories for the full-length proteins. The identification of proteolytic fragments was less consistent, being affected by different fixation and staining methods. Assay pH did not influence the time to disappearance of the full-length protein or protein fragments, however, results across laboratories were more consistent at pH 1.2 than at pH 2.0 [15].

GM crops have great biotechnological potential. A rigorous safety assessment tree has been established for assessing the safety of GM crops. Currently, no single factor is recognized as an identifier for protein allergenicity [19]. The stability of a protein to digestion, as predicted by an in vitro SGF assay, currently is used as one element in the risk assessment process. In vitro gastrointestinal digestion protocols should be preferably combined with immunological assays in order to elucidate the role of large digestion-resistant fragments and the influence of the food matrix on the stimulation of the immune system [55]. There is also a need to establish standardized assay conditions so that direct comparison of results from different laboratories can be made. Consensus also needs to be reached on relating the measured digestibility to the allergenic potential of proteins [56]. Experiments carried out by digesting purified proteins do not take into account the effect of food matrix and food processing on proteolytic activity and food protein digestion, and therefore the digestibility predictions for these proteins may be misleading for proteins in certain matrices [54].

Ofori-Anti et al. [54] tried to establish objective detection limits for the pepsin digestion assay that may be used in the assessment of GM foods. They proposed assay conditions in which 10 U of pepsin will be added per microgram of test protein and end point of digestion will be assessed after 20 min (Table 8.1). Based on their observations, proteins that are not degraded by 90% within 20 min of digestion carry an increased risk of causing food allergies. In addition, proteins that produce pepsin-stable peptide fragments of more than 5 kDa may carry an increased risk of causing food allergies [54].

A simplified assay used for digestibility assessment of novel proteins, although easy to apply and standardize, has often been criticized for not being physiologically relevant. Physiologically relevant digestion protocols, however, should take into account many different factors, and validating such a digestion protocol still represents a challenging task for the scientific community interested in dietary protein digestion. Different factors affect digestion of food proteins during gastrointestinal digestion in the gastrointestinal tract: buffering effect of food ingredients, mechanical breakdown of food tissue, stomach pH, surfactants (phospholipids), pepsin activity, gastric lipase, emulsification of lipids, gastric emptying, etc.

The secretory capacity of the stomach changes physiologically throughout the lifetime, thus, influencing gastric protein digestion. Certain pathological conditions also affect gastric digestion. The importance of developing more relevant physiological digestion protocols is underlined by studies showing that increase of the gastric pH by antiulcer medications in patients with dyspepsia (gastritis, ulcer, erosions) represents a risk factor for food-induced allergic reactions [57].

8.4   Animal Models of Food Allergy

The potential allergenicity of newly introduced proteins in genetically engineered foods has become an important safety evaluation issue. However, to evaluate the potential allergenicity and the potency of new proteins in our food, there are still no widely accepted and reliable test systems. The best-known allergy assessment proposal for foods derived from genetically engineered plants was the stepwise process proposed by an FAO/WHO expert consultation.

For agricultural biotechnology, the safety assessment for potential protein allergenicity has two primary goals: to demonstrate that existing allergens, or likely cross-reactive proteins, are not transferred from the recognized source of the gene into a food crop and to demonstrate that an introduced protein is unlikely to become a food allergen *de novo*. The possibility that the novel protein might become an allergen *de novo* cannot be assessed with bioinformatic methods. Instead, certain characteristics attributed to common food allergens are evaluated, including digestibility by pepsin, the abundance of the protein in the food, and stability upon relevant processing conditions if there are concerns based on digestive stability [58].

As prediction of the sensitizing potential of the novel introduced protein based on animal testing was considered to be very important, animal models were introduced as one of the new test items, despite the fact that none of the currently studied models has been widely accepted and validated. More work is needed on identification of appropriate end points, particularly those that reflect anaphylactic activity. Before validation can be considered, decisions have to be made regarding which mouse strains and adjuvants to include, as well as the ideal doses of test materials. Appropriate test substances that represent a range from highly allergenic to poorly allergenic need to be selected [59, 60].

An appropriate animal model should produce sensitization and/or elicitation of allergic symptoms at a physiologically relevant dose, via the relevant route of exposure in a standard mouse strain (Table 8.2) [60]. A major barrier to developing such an ideal model is oral tolerance. In mice, the immune response to an ingested protein is an active process (oral tolerance) that efficiently blocks the development of IgE and delayed-type hypersensitivity responses. Although there is no doubt that parenteral (i.p.) administration of protein avoids the development of oral tolerance and provides a clear indication of the inherent ability of proteins to induce IgE anti-

Table 8.2 Main characteristics of an ideal animal model. (Reproduced with permission from Elsevier)

Use a simple protocol for sensitization and challenges by taking into account several key parameters:

Route of sensitization (e.g., oral, i.p., dermal)

Strain (e.g., in case of mice: BALB/c, C3H, A/J, BDF-1, etc.)

Adjuvant (cholera toxin, endotoxin, no adjuvant)

Biologically relevant end points (IgE, symptomatology, biological markers, cytokines)

Validity proven by dose–response curves with different sensitization and challenges

Reproducible between laboratories for measuring an allergic response

Reproducible over time (generations and seasonal)

Specific for discriminating potent food allergens from weakly allergenic to nonallergenic proteins.

Sensitive for distinguishing a threshold beyond which significant allergy would be predicted and potentially for producing graded responses comparable to what is known regarding their prevalence and severity of responses in humans, e.g., peanut > egg > the putative nonallergenic spinach

Acceptable under animal care and use protocols at an international level (e.g., www.iacuc.org)

Although it may prove useful to identify the de novo allergic potential of proteins and/or foods, to set up one animal model may not be achievable. It may be necessary to use two or more animal models for predicting the allergenic potential of proteins. If reliable models were developed, they should have the above-mentioned features [58]

body responses, it must be acknowledged that the ability of a protein to provoke an IgE antibody response, when administered parenterally, will not necessarily translate into a risk of allergic sensitization following normal dietary exposure.

So far, developed mouse models of food allergy mostly used adjuvants (such as cholera toxin and staphylococcal enterotoxin B, SEB) and the oral route of exposure [60]. Several variations of a mouse model originally developed to study peanut and cow's milk IgE-mediated allergy [61, 62] were described and recently applied to assess the relative allergenicity of various food extracts [63]. Mice were sensitized with two doses of food extract (in the 1–5 mg range), orally, 1 week apart, and specific IgE was assessed 1 week after the second dose. To circumvent oral tolerance and evoke differential IgE responses to a panel of allergenic and nonallergenic food extracts, female C3H/HeJ mice were exposed subcutaneously or orally with cholera toxin as an adjuvant. All foods elicited IgE by the subcutaneous route. Oral exposure, however, resulted in generation of IgE in response to allergens (peanut, Brazil nut, and egg white) but not to nonallergens (spinach and turkey). It should be noted that foods containing digestion-resistant proteins provoked allergic responses in this model, supporting the current use of pepsin resistance in the decision tree for potential allergenicity assessment [63].

An oral exposure model that used a much lower sensitizing dose was also described. Food antigen at the dose of 0.1 mg was given on eight consecutive days using SEB as the adjuvant [64]. SEB impaired oral tolerance and permitted allergic responses. In addition to measuring specific IgE, a challenge dose without adjuvant was administered 24 h after the last sensitization, and various physiologic and immunologic end point characteristics of anaphylaxis were assessed. SEB-driven sensitization induced eosinophilia in the blood and intestinal tissues not observed with cholera toxin sensitization. SEB impaired tolerance specifically by impairing expression of TGF-beta and regulatory T cells, and tolerance was restored with high-dose antigen. This model gave promising results for peanut extract and ovalbumin.

A model that sensitizes mice to food extracts of hazelnut and cashew nuts by transdermal exposure was described [65–67]. BALB/c mice were repeatedly exposed to adjuvant-free hazelnut protein via the transdermal route and systemic allergic and anaphylactic responses were studied. Transdermal exposure to hazelnut protein elicited a robust systemic IgE response in a dose-dependent manner. Oral challenge of transdermally sensitized mice with hazelnut protein resulted in immediate clinical signs of systemic anaphylaxis. Clinical hypersensitivity reaction was associated with severe pathological changes in the small intestine [65].

Another murine model assessed the induction of oral tolerance rather than sensitization, because an allergic response requires the ability to sensitize, as well as to avoid the induction of oral tolerance. Mice were fed a single dose of vehicle or food extract without adjuvant and challenged 1 week later by the i.p. route with the food extract of interest. Significantly lower IgE levels in mice that received the oral food extract exposure versus vehicle were considered to be indicative of oral tolerance, which mitigates allergenicity [63]. In their previous work, authors demonstrated that pepsin resistance is important for sensitization induction in a murine model of food allergy [63], while resistance to both pepsin and trypsin appears to be required for oral tolerance induction in the murine model of oral tolerance [68]. Altering digestibility, pH, and/or solubility of the sensitizing food extract can change the results obtained in the oral animal models, suggesting that the food matrix can affect results [69, 70]. The role of the food matrix in food allergy induction requires further study. It has been reported that purified peanut allergens possess little intrinsic immune-stimulating capacity in contrast to a whole peanut extract [71]. The data indicate that the food matrix can influence responses to individual proteins and, therefore, the food matrix should be taken into account when developing models for allergenic potential assessment.

Furthermore, recent study aimed at validating a mouse model for cow's milk allergy to assess the potential allergenicity of hydrolyzed cow's milk-based infant formulas. According to the European Commission directive 2006/141/E on infant formulas, the hypoallergenicity of hydrolyzed infant formulas needs to be assessed by showing that the hypoallergenic formulas are not able to sensitize animals to the protein source they are derived from. The transferability and the discriminatory power of a mouse model were evaluated in four research centers. Mice were sensitized by oral gavage with whey, or extensively hydrolyzed whey, using cholera toxin as an adjuvant. The cow's milk allergy mouse model was capable to distinguish the sensitizing capacity of complete and hydrolyzed cow's milk protein. The model can be effectively transferred between different laboratories. The authors proposed this mouse model as a new strategy for the screening of new hypoallergenic cow's milk formulas [72].

The rat model of food allergy has also been described [73–75]. Young Brown Norway rats were exposed to 1 mg ovalbumin by daily gavage dosing for 42 days without the use of an adjuvant. Ovalbumin-specific IgE and IgG responses were determined by enzyme-linked immunosorbent assay (ELISA). On an oral challenge with ovalbumin, some clinical symptoms of food allergy-like effects on the respiratory system, blood pressure, and permeability of the gastrointestinal barrier were studied. In addition, rats were orally exposed to a total hen egg white protein extract and cow's milk, and the specificities of induced antibody responses were compared with the specificities of antibodies in sera from egg- and milk-allergic patients using immunoblotting. Animals orally exposed to the allergens developed specific IgE and IgG antibodies which recognized the same proteins compared with antibodies from egg- or cow's milk-allergic patients. Among the various clinical symptoms of food allergy, gut permeability was increased after an oral challenge. These data support the suitability of the rat animal model for food allergy research and for the study of the allergenicity of novel food proteins [74].

References

1. Miller SA, Artuso A, Avery D, Beachy RN, Day PR, Fennema OR, Hardy R, Keeling PL, Klaenhammer TR, McGloughlin M *et al*: **Benefits and concerns associated with recombinant DNA biotechnology-derived foods**. *Food Technol* 2000, **54**(10):61.
2. Metcalfe DD: **Genetically modified crops and allergenicity**. *Nat Immunol* 2005, **6**(9):857–860.
3. Harlander SK: **The evolution of modern agriculture and its future with biotechnology**. *J Am Coll Nutr* 2002, **21**(3 Suppl):161S–165S.
4. Nordlee JA, Taylor SL, Townsend JA, Thomas LA, Bush RK: **Identification of a Brazil-nut allergen in transgenic soybeans**. *N Engl J Med* 1996, **334**(11):688–692.
5. European Food Safety Authority (EFSA). **Guidance for risk assessment of food and feed from genetically modified plants**. *EFSA Journal* 2011, **9**(5):2150
6. Codex Alimentarius Commission. Alinorm 03/34: Joint FAO/WHO Food Standard Programme CAC, Twenty-Fifth Session, Rome, 30 June–5 July, 2003: **Appendix III, Guideline for the conduct of food safety assessment of foods derived from recombinant-DNA plants and Appendix IV, Annex on the assessment of possible allergenicity**. 2003:47–60.
7. Codex Alimentarius Commission: **Foods Derived From Modern Biotechnology**. *FAO/WHO, Rome* 2009:1–85.
8. Goodman RE, Vieths S, Sampson HA, Hill D, Ebisawa M, Taylor SL, van Ree R: **Allergenicity assessment of genetically modified crops—what makes sense?** *Nature Biotechol* 2008, **26**(1):73–81.
9. Metcalfe DD, Astwood JD, Townsend R, Sampson HA, Taylor SL, Fuchs RL: **Assessment of the allergenic potential of foods derived from genetically engineered crop plants**. *Crit Rev Food Sci Nutr* 1996, **36 Suppl**:S165–186.
10. FAO/WHO: **Evaluation of allergenicity of genetically modified foods**. *Report of a joint FAO/WHO expert consultation on allergenicity of foods derived from biotechnology (Food and Agriculture Organization of the United Nations (FAO), Rome, 2001.*
11. van Ree R: **Carbohydrate epitopes and their relevance for the diagnosis and treatment of allergic diseases**. *Int Arch Allergy Immunol* 2002, **129**(3):189–197.
12. Altmann F: **The role of protein glycosylation in allergy**. *Int Arch Allergy Immunol* 2007, **142**(2):99–115.
13. Asero R, Mistrello G, Roncarolo D, de Vries SC, Gautier MF, Ciurana CL, Verbeek E, Mohammadi T, Knul-Brettlova V, Akkerdaas JH *et al*: **Lipid transfer protein: a pan-allergen**

in plant-derived foods that is highly resistant to pepsin digestion. *Int Arch Allergy Immunol* 2000, **122**(1):20–32.

14. Fu TT, Abbott UR, Hatzos C: **Digestibility of food allergens and nonallergenic proteins in simulated gastric fluid and simulated intestinal fluid—A comparative study.** *J Agric Food Chem* 2002, **50**(24):7154–7160.

15. Thomas K, Aalbers M, Bannon GA, Bartels M, Dearman RJ, Esdaile DJ, Fu TJ, Glatt CM, Hadfield N, Hatzos C *et al*: **A multi-laboratory evaluation of a common in vitro pepsin digestion assay protocol used in assessing the safety of novel proteins.** *Regul Toxicol Pharmacol* 2004, **39**(2):87–98.

16. EFSA Panel on Genetically Modified Organisms (GMO): **Scientific Opinion on the assessment of allergenicity of GM plants and microorganisms and derived food and feed.** *EFSA Journal* 2010, **8**(7):168.

17. Hagen JB: **The origins of bioinformatics.** *Nat Rev Genet* 2000, **1**(3):231–236.

18. Baxevanis AD, Ouellette B (eds.): **Bioinformatics: A practical guide to the analysis of genee and proteins:** John Wiley & Sons, New Jersey 2005.

19. Ladics GS: **Current codex guidelines for assessment of potential protein allergenicity.** *Food Chem Toxicol* 2008, **46 Suppl 10**:S20–23.

20. Ivanciuc O, Garcia T, Torres M, Schein CH, Braun W: **Characteristic motifs for families of allergenic proteins.** *Mol Immunol* 2009, **46**(4):559–568.

21. Radauer C, Bublin M, Wagner S, Mari A, Breiteneder H: **Allergens are distributed into few protein families and possess a restricted number of biochemical functions.** *J Allergy Clin Immunol* 2008, **121**(4):847–852 e847.

22. Gendel SM: **Sequence analysis for assessing potential allergenicity.** *Ann N Y Acad Sci* 2002, **964**:87–98.

23. Stadler MB, Stadler BM: **Allergenicity prediction by protein sequence.** *Faseb J* 2003, **17**(9):1141–1143.

24. Ladics GS, Selgrade MK: **Identifying food proteins with allergenic potential: evolution of approaches to safety assessment and research to provide additional tools.** *Regul Toxicol Pharmacol* 2009, **54**(3 Suppl):S2–6.

25. Bailey TL, Elkan C: **Fitting a mixture model by expectation-maximization to discover motifs in biopolymers.** In: *Second International Conference on Intelligent Systems for Molecular Biology Menlo Park, CA: AAAI Press: 1994*; 1994: 28–36.

26. Li KB, Issac P, Krishnan A: **Predicting allergenic proteins using wavelet transform.** *Bioinformatics* 2004, **20**(16):2572–2578.

27. Bjorklund AK, Soeria-Atmadja D, Zorzet A, Hammerling U, Gustafsson MG: **Supervised identification of allergen-representative peptides for in silico detection of potentially allergenic proteins.** *Bioinformatics* 2005, **21**(1):39–50.

28. Saha S, Raghava GP: **AlgPred: prediction of allergenic proteins and mapping of IgE epitopes.** *Nucleic Acids Res* 2006, **34**(Web Server issue):W202–209.

29. Bailey TL, Gribskov M: **Score distributions for simultaneous matching to multiple motifs.** *J Comput Biol* 1997, **4**(1):45–59.

30. Wold S, Jonsson J, Sjostrom M, Sandberg M, Rannar S: **DNA and peptide sequences and chemical processes multivariately modelled by principal component analysis and partial least-squares projections to latent structures.** *Anal Chim Acta* 1993, **277**(2):239–253.

31. Nyström A, Andersson PM, Lundstedt T: **Multivariate data analysis of topographically modified α-melanotropin analogues using auto and cross auto covariances (ACC).** *Quant Struct-Act Relat* 2000, **19**(3):264–269.

32. Lapinsh M, Gutcaits A, Prusis P, Post C, Lundstedt T, Wikberg JE: **Classification of G-protein coupled receptors by alignment-independent extraction of principal chemical properties of primary amino acid sequences.** *Protein Sci* 2002, **11**(4):795–805.

33. Dimitrov I, Flower DR, Doytchinova I: **AllerTOP–a server for in silico prediction of allergens.** *BMC Bioinformatics* 2013, **14 Suppl 6**:S4.

34. Gendel SM: **Allergen databases and allergen semantics.** *Regul Toxicol Pharmacol* 2009, **54**(3 Suppl):S7–10.

35. King TP, Hoffman D, Lowenstein H, Marsh DG, Platts-Mills TA, Thomas W: **Allergen no-menclature. WHO/IUIS Allergen Nomenclature Subcommittee.** *Int Arch Allergy Immu-nol* 1994, **105**(3):224–233.

36. Mari A, Scala E, Palazzo P, Ridolfi S, Zennaro D, Carabella G: **Bioinformatics applied to allergy: allergen databases, from collecting sequence information to data integration. The Allergome platform as a model.** *Cell Immunol* 2006, **244**(2):97–100.

37. Soeria-Atmadja D, Zorzet A, Gustafsson MG, Hammerling U: **Statistical evaluation of local alignment features predicting allergenicity using supervised classification algorithms.** *Int Arch Allergy Immunol* 2004, **133**(2):101–112.

38. King TP, Hoffman D, Lowenstein H, Marsh DG, Platts-Mills TAE, Thomas W: **Allergen nomenclature.** *Bulletin of the World Health Organization* 1994, **72**(5):797–806.

39. King TP, Hoffman D, Lowenstein H, Marsh DG, Platts-Mills TA, Thomas W: **Allergen no-menclature.** *Allergy* 1995, **50**(9):765–774.

40. Chapman MD, Pomes A, Breiteneder H, Ferreira F: **Nomenclature and structural biology of allergens.** *J Allergy Clin Immunol* 2007, **119**(2):414–420.

41. Boutet E, Lieberherr D, Tognolli M, Schneider M, Bairoch A: **UniProtKB/Swiss-Prot.** *Methods Mol Biol* 2007, **406**:89–112.

42. Westbrook J, Feng Z, Jain S, Bhat TN, Thanki N, Ravichandran V, Gilliland GL, Bluhm W, Weissig H, Greer DS et al: **The Protein Data Bank: unifying the archive.** *Nucleic Acids Res* 2002, **30**(1):245–248.

43. Sayers EW, Barrett T, Benson DA, Bolton E, Bryant SH, Canese K, Chetvernin V, Church DM, Dicuccio M, Federhen S et al: **Database resources of the National Center for Bio-technology Information.** *Nucleic Acids Res* 2012, **40**(Database issue):D13–25.

44. Hileman RE, Silvanovich A, Goodman RE, Rice EA, Holleschak G, Astwood JD, Hefle SL: **Bioinformatic methods for allergenicity assessment using a comprehensive allergen da-tabase.** *Int Arch Allergy Immunol* 2002, **128**(4):280–291.

45. Ivanciuc O, Schein CH, Braun W: **SDAP: database and computational tools for allergenic proteins.** *Nucleic Acids Res* 2003, **31**(1):359–362.

46. Bairoch A, Apweiler R: **The SWISS-PROT protein sequence database and its supple-ment TrEMBL in 2000.** *Nucleic Acids Res* 2000, **28**(1):45–48.

47. Barker WC, Garavelli JS, Hou Z, Huang H, Ledley RS, McGarvey PB, Mewes HW, Orcutt BC, Pfeiffer F, Tsugita A et al: **Protein Information Resource: a community resource for expert annotation of protein data.** *Nucleic Acids Res* 2001, **29**(1):29–32.

48. Wheeler DL, Church DM, Lash AE, Leipe DD, Madden TL, Pontius JU, Schuler GD, Schriml LM, Tatusova TA, Wagner L et al: **Database resources of the National Center for Biotechnology Information.** *Nucleic Acids Res* 2001, **29**(1):11–16.

49. Berman HM, Westbrook J, Feng Z, Gilliland G, Bhat TN, Weissig H, Shindyalov IN, Bourne PE: **The Protein Data Bank.** *Nucleic Acids Res* 2000, **28**(1):235–242.

50. Fiers MW, Kleter GA, Nijland H, Peijnenburg AA, Nap JP, van Ham RC: **Allermatch, a webtool for the prediction of potential allergenicity according to current FAO/WHO Codex alimentarius guidelines.** *BMC Bioinformatics* 2004, **5**:133.

51. Zhang ZH, Koh JL, Zhang GL, Choo KH, Tammi MT, Tong JC: **AllerTool: a web server for predicting allergenicity and allergic cross-reactivity in proteins.** *Bioinformatics* 2007, **23**(4):504–506.

52. Thomas K, Herouet-Guicheney C, Ladics G, Bannon G, Cockburn A, Crevel R, Fitzpatrick J, Mills C, Privalle L, Vieths S: **Evaluating the effect of food processing on the potential hu-man allergenicity of novel proteins: international workshop report.** *Food Chem Toxicol* 2007, **45**(7):1116–1122.

53. Astwood JD, Leach JN, Fuchs RL: **Stability of food allergens to digestion in vitro.** *Nat Biotechnol* 1996, **14**(10):1269–1273.

54. Ofori-Anti AO, Ariyarathna H, Chen L, Lee HL, Pramod SN, Goodman RE: **Establishing objective detection limits for the pepsin digestion assay used in the assessment of geneti-cally modified foods.** *Regul Toxicol Pharmacol* 2008, **52**(2):94–103.

55. Moreno FJ: **Gastrointestinal digestion of food allergens: effect on their allergenicity.** *Biomed Pharmacother* 2007, **61**(1):50–60.

56. Fu TJ: **Digestion stability as a criterion for protein allergenicity assessment.** *Ann N Y Acad Sci* 2002, **964**:99–110.
57. Untersmayr E, Jensen-Jarolim E: **The role of protein digestibility and antacids on food allergy outcomes.** *J Allergy Clin Immunol* 2008, **121**(6):1301–1308; quiz 1309–1310.
58. Ladics GS, Knippels LM, Penninks AH, Bannon GA, Goodman RE, Herouet-Guicheney C: **Review of animal models designed to predict the potential allergenicity of novel proteins in genetically modified crops.** *Regul Toxicol Pharmacol* 2010, **56**(2):212–224.
59. Atherton KT, Dearman RJ, Kimber I: **Protein allergenicity in mice: a potential approach for hazard identification.** *Ann N Y Acad Sci* 2002, **964**:163–171.
60. Selgrade MK, Bowman CC, Ladics GS, Privalle L, Laessig SA: **Safety assessment of biotechnology products for potential risk of food allergy: implications of new research.** *Toxicol Sci* 2009, **110**(1):31–39.
61. Li XM, Schofield BH, Huang CK, Kleiner GI, Sampson HA: **A murine model of IgE-mediated cow's milk hypersensitivity.** *J Allergy Clin Immunol* 1999, **103**(2 Pt 1):206–214.
62. Li XM, Serebrisky D, Lee SY, Huang CK, Bardina L, Schofield BH, Stanley JS, Burks AW, Bannon GA, Sampson HA: **A murine model of peanut anaphylaxis: T- and B-cell responses to a major peanut allergen mimic human responses.** *J Allergy Clin Immunol* 2000, **106**(1 Pt 1):150–158.
63. Bowman CC, Selgrade MK: **Differences in allergenic potential of food extracts following oral exposure in mice reflect differences in digestibility: potential approaches to safety assessment.** *Toxicol Sci* 2008, **102**(1):100–109.
64. Ganeshan K, Neilsen CV, Hadsaitong A, Schleimer RP, Luo X, Bryce PJ: **Impairing oral tolerance promotes allergy and anaphylaxis: a new murine food allergy model.** *J Allergy Clin Immunol* 2009, **123**(1):231–238 e234.
65. Birmingham NP, Parvataneni S, Hassan HM, Harkema J, Samineni S, Navuluri L, Kelly CJ, Gangur V: **An adjuvant-free mouse model of tree nut allergy using hazelnut as a model tree nut.** *Int Arch Allergy Immunol* 2007, **144**(3):203–210.
66. Gonipeta B, Parvataneni S, Paruchuri P, Gangur V: **Long-term characteristics of hazelnut allergy in an adjuvant-free mouse model.** *Int Arch Allergy Immunol* 2010, **152**(3):219–225.
67. Parvataneni S, Gonipeta B, Tempelman RJ, Gangur V: **Development of an adjuvant-free cashew nut allergy mouse model.** *Int Arch Allergy Immunol* 2009, **149**(4):299–304.
68. Bowman CC, Selgrade MK: **Failure to induce oral tolerance in mice is predictive of dietary allergenic potency among foods with sensitizing capacity.** *Toxicol Sci* 2008, **106**(2):435–443.
69. Thomas K, MacIntosh S, Bannon G, Herouet-Guicheney C, Holsapple M, Ladics G, McClain S, Vieths S, Woolhiser M, Privalle L: **Scientific advancement of novel protein allergenicity evaluation: an overview of work from the HESI Protein Allergenicity Technical Committee (2000–2008).** *Food Chem Toxicol* 2009, **47**(6):1041–1050.
70. Mills EN, Mackie AR: **The impact of processing on allergenicity of food.** *Curr Opin Allergy Clin Immunol* 2008, **8**(3):249–253.
71. van Wijk F, Nierkens S, Hassing I, Feijen M, Koppelman SJ, de Jong GA, Pieters R, Knippels LM: **The effect of the food matrix on in vivo immune responses to purified peanut allergens.** *Toxicol Sci* 2005, **86**(2):333–341.
72. van Esch BC, van Bilsen JH, Jeurink PV, Garssen J, Penninks AH, Smit JJ, Pieters RH, Knippels LM: **Interlaboratory evaluation of a cow's milk allergy mouse model to assess the allergenicity of hydrolysed cow's milk based infant formulas.** *Toxicol Lett* 2013, **220**(1):95–102.
73. Knippels LM, Houben GF, Spanhaak S, Penninks AH: **An oral sensitization model in Brown Norway rats to screen for potential allergenicity of food proteins.** *Methods* 1999, **19**(1):78–82.
74. Knippels LM, Penninks AH: **Assessment of protein allergenicity: studies in brown norway rats.** *Ann N Y Acad Sci* 2002, **964**:151–161.
75. Schouten B, van Esch BC, Hofman GA, de Kivit S, Boon L, Knippels LM, Garssen J, Willemsen LE: **A potential role for CD25 + regulatory T-cells in the protection against casein allergy by dietary non-digestible carbohydrates.** *Br J Nutr* 2012, **107**(1):96–105.

Index

T. Ćirković Veličković, M. Gavrović-Jankulović, *Food Allergens,*
Food Microbiology and Food Safety, DOI 10.1007/978-1-4939-0841-7,
© Springer Science+Business Media New York 2014

Printed by Printforce, the Netherlands